Anticandidal T

Anticandidal Therapeutics

Discovery and Development

Awanish Kumar
Associate Professor, Department of Biotechnology,
National Institute of Technology Raipur, India

Anubhuti Jha
Ph.D. Scholar, Department of Biotechnology,
National Institute of Technology Raipur, India

ELSEVIER

ACADEMIC PRESS
An imprint of Elsevier

Academic Press is an imprint of Elsevier
125 London Wall, London EC2Y 5AS, United Kingdom
525 B Street, Suite 1650, San Diego, CA 92101, United States
50 Hampshire Street, 5th Floor, Cambridge, MA 02139, United States
The Boulevard, Langford Lane, Kidlington, Oxford OX5 1GB, United Kingdom

Copyright © 2023 Elsevier Inc. All rights reserved.

No part of this publication may be reproduced or transmitted in any form or by any means, electronic or mechanical, including photocopying, recording, or any information storage and retrieval system, without permission in writing from the publisher. Details on how to seek permission, further information about the Publisher's permissions policies and our arrangements with organizations such as the Copyright Clearance Center and the Copyright Licensing Agency, can be found at our website: www.elsevier.com/permissions.

This book and the individual contributions contained in it are protected under copyright by the Publisher (other than as may be noted herein).

Notices

Knowledge and best practice in this field are constantly changing. As new research and experience broaden our understanding, changes in research methods, professional practices, or medical treatment may become necessary.

Practitioners and researchers must always rely on their own experience and knowledge in evaluating and using any information, methods, compounds, or experiments described herein. In using such information or methods they should be mindful of their own safety and the safety of others, including parties for whom they have a professional responsibility.

To the fullest extent of the law, neither the Publisher nor the authors, contributors, or editors, assume any liability for any injury and/or damage to persons or property as a matter of products liability, negligence or otherwise, or from any use or operation of any methods, products, instructions, or ideas contained in the material herein.

ISBN: 978-0-443-18744-5

For information on all Academic Press publications visit our website at
https://www.elsevier.com/books-and-journals

Publisher: Stacy Masucci
Acquisitions Editor: Linda Versteeg-Buschman
Editorial Project Manager: Michaela Realiza
Production Project Manager: Kiruthika Govindaraju
Cover Designer: Christian J. Bilbow

Typeset by TNQ Technologies

Dedication

We dedicate this book to people impacted by the book and the readers of the book.

Contents

PREFACE xiii
ACKNOWLEDGMENTS xv

CHAPTER 1 History of infection 1
    Introduction 1
    Introduction to plague 2
    Types of plague 3
    14th century plagues 4
    Plague of Athens 5
    Second plague pandemic 9
    Black Death 9
    Third plague pandemic 12
    Smallpox 13
    Tuberculosis 14
    Spanish flu 16
    HIV/AIDS 18
    Swine flu 21
    COVID-19 (current pandemic) 21
    A note on the future 22
    References 25

CHAPTER 2 Brief introduction on infection 27
    Introduction 27
    References 31

CHAPTER 3 Fungal diseases and antifungal drugs 33
    Introduction 33
    Superficial mycoses 33
    Cutaneous mycoses 34
    Subcutaneous mycoses 34
    Systemic or invasive mycoses 35
    Candidal infection 38
    Biofilm formation 39

	Adhesion and invasins	39
	Contact sensing and thigmotropism	40
	Hydrolytic enzymes	41
	pH sensing and regulation	41
	Metabolic adaptation	43
	Environmental stress response	43
	Heat shock protein	44
	Metal acquisition	45
	Antifungal agents	45
	Azoles	45
	Polyenes	46
	Echinocandins	47
	Allylamines	48
	Other drugs	48
	Mechanism of antifungal resistance	49
	Resistance to azoles	49
	Alteration of drug target	50
	Upregulation of multidrug transporters	50
	Cellular stress responses	51
	Biofilms	52
	Modification of ergosterol biosynthetic pathway	52
	Chromosomal alterations	52
	Resistance to polyenes	52
	Resistance to echinocandins	53
	Upregulation of multidrug transporters	53
	Drug efflux	54
	References	55
CHAPTER 4	Multidrug transporters of fungal pathogen Candida	65
	Introduction	65
	ABC transporters	65
	MFS transporters	66
	Regulation of MDR	66
	PDR1 and *PDR3* are xenobiotic receptor in *S. cerevisiae*	67
	Xenobiotic regulation	68
	Efflux pumps regulation	70
	References	72
CHAPTER 5	Essential anticandidal targets	77
	Fungal cell wall	77
	Fungal membrane components	78
	Inhibition of heat shock protein 90	79
	Inhibition of calcineurin signaling	81

	Signaling pathways	81
	Cell cycle control pathways	82
	Membrane transporters	83
	Other surface targets	87
	Cross-talk	88
	References	89
CHAPTER 6	**Stages of anticandidal agent development**	**97**
	Introduction	97
	Drug target discovery	99
	Screening	99
	HIT identification	100
	Lead optimization	100
	Experimentation	101
	Characterization of product	101
	Formulation, delivery, packaging development	102
	Pharmacokinetics and drug disposition	102
	Preclinical testing and IND application	102
	Acute studies	103
	Repeated dose studies	103
	Generic toxicity studies	104
	Reproductive toxicity studies	104
	Carcinogenicity studies	104
	Toxicokinetic studies	104
	Bioanalytical testing	105
	Clinical trials	105
	Postmarketing surveillance	106
	References	107
CHAPTER 7	**Recent trends and progress in antifungal translational research**	**111**
	Introduction	111
	Translational research and strategy in fungal infection	112
	References	116
CHAPTER 8	**Clinical status of anticandidal therapeutic agents**	**119**
	Introduction	119
	Synthetic compounds as futuristic antifungal	119
	Olorofim: orotomides (F901318)	120
	Natural compounds as promising antifungal	120
	Curcumin	121
	Thymol	122

Berberine 123
Tetrandrine 124
Antifungals in different stage of clinical trials 125
Rezafungin 125
Ibrexafungerp (SCY-078) 125
Fosmanogepix (APX001) 126
Nikkomycin Z 126
Tetrazoles 127
Encochleated amphotericin B (MAT2203) 127
F901318 (olorofim) 129
VL-2397 129
T-2307 130
AR-12 130
References 131

CHAPTER 9 Drug repurposing for development of effective anticandidals 137
Introduction 137
References 144

CHAPTER 10 Commercial aspects of drug development 147
Introduction 147
Drug development process 148
The design of a clinical trial 151
Different types of clinical research 151
Informed consent 153
The investigational new drug process 153
Drug review 154
New drug application 154
Postmarketing drug safety monitoring 155
Generic drugs 155
Reasons for catastrophic marketing failures 156
Failures of pharmaceutical companies 160
References 161

CHAPTER 11 Challenges and future prospects 163
Introduction 163
Gray areas in drug discovery research 164
Cost of new drug development 166
Out-of-pocket costs 169
Success rates 169
Development timelines 169
Cost of capital 170

  Preclinical phase 170
  Clinical trials phase 170
  Macroeconomic, demographic, and policy factors
  affecting R&D funding 174
  References 175

INDEX..177

Preface

Below are a few lines about the book *Anticandidal Therapeutics*, which would be the crux of the book. This book is written with the intent to provide valuable information on antifungal therapeutics with special emphasis on the anticandidal agent. Authors are working in this area. The various chapters discussed in this book collectively address the important topics and update material on these topics. This is an attempt to expand the perspective of the reader on the complex problem of antifungal resistance, challenges, and development of future therapeutics. Chapters contain various important topics like fungal diseases and antifungal drugs, multidrug transporters of the fungal pathogen *Candida*, essential anticandidal targets, stages of anticandidal agent development, recent trends and progress in antifungal translational research, clinical studies status of anticandidal therapeutic agents, drug repurposing for development of effective anticandidals, commercial aspects of drug development, challenges, and future prospects. The content of the book targets the broad readership of academicians, college students, research staff, and investigators working in this specific area.

Acknowledgments

We are overwhelmed in all humbleness and gratefulness to acknowledge our depth to all those who have helped us to put these ideas in the form of the book.

At the onset, we would like to express our special thanks of gratitude to our organization "National Institute of Technology Raipur" (India).

We thank our colleagues, friends, and research scholars who have helped directly or indirectly in completing this task.

Completion of this project could not have been accomplished without the support of our family members. I, Dr. Awanish Kumar, especially thank my caring, loving, and supportive wife *Smriti*, and my sweet and affectionate daughter *Anika*. I am eternally grateful to my father *Shri Anugrah Narayan Singh* and my mother *Mrs. Saroj Devi* for their blessings and for making me what I am today.

Our heartfelt thanks to the publishing team of Elsevier.

At last, we bow our heads before God Almighty for his blessings in the successful and timely completion of this book.

Dr. Awanish Kumar

Dr. Anubhuti Jha

CHAPTER 1

History of infection

Introduction

According to the Spanish philosopher George Santayana, "Those who don't know history are destined to repeat it." One should never forget where they came from. The history of infectious diseases warns us of the same notion more starkly—sometimes trying to make us aware beforehand, other times, making us afraid so much that we tend to be visibly scared every time we breathe, but most importantly, it guides and prepares us for a future full of uncertainties, vulnerabilities, and also, possibilities. Without a thorough understanding of how infectious diseases came to be, survived to thrive, ruled to abolish, and remained deadly to create havoc beyond one's imagination, it is futile to discuss how they can be tackled, managed, and dealt with effectively. Since the beginning of humanity on Earth, infectious diseases have affected the way humans have interacted with each other and nature. To understand and tackle this issue, it becomes crucial to comprehend the origin, spread, function and casualty these diseases created. In order to deal with the disease effectively, we must inspect the cause and effect deeper and with precision. Studying the history of infectious diseases can be the first step toward a long journey. The historical knowledge has been retrieved thanks to the historical information and vivid description of the infectious diseases individual historians who witnessed the brutality of the disease themselves have documented. Many of these historians' accuracy has been confirmed by modern researchers with empirical data and excavations. A majority of plagues seem to receive a perfect ground to grow and expand near ports with heavy human activity, poor hygiene, and close proximity. These epidemics are also notorious for returning with much force and increased infections, as the interval between one outbreak and

CONTENTS

Introduction 1
Introduction to plague 2
Types of plague 3
14th century plagues 4
Plague of Athens 5
Second plague pandemic 9
Black Death 9
Third plague pandemic 12
Smallpox 13
Tuberculosis .. 14
Spanish flu 16
HIV/AIDS 18
Swine flu 21

Anticandidal Therapeutics. https://doi.org/10.1016/B978-0-443-18744-5.00010-1
Copyright © 2023 Elsevier Inc. All rights reserved.

another does not provide enough time for recovery and proper medical preparation.

With each infectious disease, people started discovering more advanced methods to prevent and diagnose them. This led to an advancement in public health and hygiene culture. Microbiologists started to understand the functionalities of bacteria, virus, fungi, protozoa, and parasites better. Working on the pathogens, the advancement on the treatment, prevention, and control have been immense. One of the profound impacts infectious diseases had on the human population was mass migration. Civilization after civilizations have been destroyed and rebuilt due to the death toll and economic burden a pandemic puts on the human population. The idea that infectious disease history has everything to do with human history, impacting it more profoundly and severely than world wars, should not be surprising [1].

Advancement in traveling led to advancement in the spread of infectious diseases. For example, European travelers between 1518 and 1568 introduced smallpox, measles, and typhus to the Central and Latin American native population. Smallpox alone was responsible for wiping out 17 million of the population in Mexico, making the population to be just under 3 million [2].

Introduction to plague

Plague is caused by a zoonotic bacteria, *Yersinia pestis*, found in small mammals and their fleas. The patients of plague start to develop symptoms in 1–7 days of getting infected [3]. It can spread directly when the infected tissue is in contact, by the bite of an infection-carrying flea, being in contact with contaminated material, and by inhaling the infected droplets. The bacteria can be easily destroyed when exposed to sunlight but it can still survive for up to 1 h in favorable conditions. When left untreated, the plague can prove to be fatal with a fatality rate of 30%–60% and in some cases death being the only option. With the help of antibiotics, the patients can be treated after an early diagnosis. In medieval times, plague occurred pretty frequently because of poor hygiene and lack of proper treatment. Modern treatments have given us a ground to have a chance to treat the patient but that does not mean that we have completely eliminated the plague-causing bacteria itself. With time, the pathogen evolves too with far greater capabilities to survive in the host for a considerable period of time, so much that it can affect the daily functioning of the patient and can infect the people surrounding them. According to the World Health Organization (WHO), during the period of 2010−2015, 3248 cases of plague were reported, resulting in 584 deaths [4].

Types of plague

1. Bubonic plague: It is the most common form of plague, caused by *Y. pestis* when an infected flea bites a person. Reaching inside the skin, the bacteria replicates in the lymph node and causes headache, fever, weakness, and chills in the patient. An inflamed lymph node, also known as "bubo," at an advanced stage, can become an open sore and release pus. It is a rarity for this infection to transmit from human to human directly. Once it reaches the lungs, the disease displays its severity.
2. Pneumonic plague: It is the most severe and virulent of all the types of plague with an incubation period as short as 24 h. Not only does it cause grave discomfort in breathing, but it also spreads from person to person through the air. An infected droplet when inhaled can infect a healthy person when there is proximity with the patients. Untreated bubonic plague can also turn into pneumonic plague when the bacteria enter the lungs. If diagnosed within 24 h of the symptom, recovery rate increases.
3. Septicemic plague: When the bacteria start multiplying in the blood, without showing any open sore, this plague occurs. All the other symptoms are the same as bubonic plague. It does not spread from person to person.

To address the present problems, one must study and analyze the historical data, their shortcomings, and progression.

14th century plagues

There are some speculations about the origin and spread of the epidemic. According to a modern theory about the spread of the disease, the climate change in the Asia region led the rodents to shift from grasslands to more populated areas. *Y. pestis* is commonly present in fleas carried by ground rodents. Discovered in 1894 in Hong Kong by Alexandre Yersin, it was suggested that rodents can be the carrier of the plague. 4 years later, Paul-Louis Simond found that fleas were also involved in the transmission as the bacteria replicated itself in their midgut, blocking it, leading the fleas to starve. This starvation leads to an aggressive food consumption behavior, but due to the blockage, the bacteria are thrown to the feeding ground by a process called regurgitation. This results in multiple numbers of bacteria infecting the host.

It should be noted that the bacteria survive and thrive with two survival strategies. First is, it stays in the host which is resistant to the disease which maintains the epidemic status quo. Second, it kills the nonresistant hosts gradually, and once done, it moves on to the resistant host. Polymerase Chain Reaction (PCR) technique has confirmed the role of *Y. pestis* present in the bodies from the mass excavation done in Asia and Europe. However, some researchers also suggest that fleas and rodents are not the only mechanism by which the disease can be spread. A large range of animals might have been involved to carry out epidemics of this scale.

Genetic analysis from mass graves suggest that the origin of the bacteria could be near the China and Kyrgyzstan border. The theory speculates that the disease might have traveled through the Silk Road which was used by Mongol invaders. Before it reached Constantinople in 1347, it had already wiped out 25 million of the population in Asia in 15 years. However, there is no historical evidence that it affected India and China severely. It created havoc in European countries, Western Asia, and Northern African countries.

The symptoms for bubonic plague included fever, headache, nausea, vomiting of blood, and joint pain. If there is no treatment available for the patient, around 80% of the patients die within 8 days. Pneumonic plague affects the respiratory system, causing cough and fever with a mortality rate of 90%–95% [5].

The plague affected the society economically, as the prices soared, environmentally, as the pandemic caused the climate to cool down by freezing up land and increasing reforestation, religiously, as the cases of persecutions based on superstitions and religious fanaticism and socially, as it brought Renaissance in Europe, following the fall of Byzantine Empire and the city of Florence. This was the same period when the word "quarantine" was coined in which the people who arrived from plague-affected regions had to isolate themselves for 40 days.

Plague of Athens

It became the worst of times when in the second year of the Peloponnesian War between Athens and Sparta in 430 BCE, a plague outbreak disrupted its well-built societies into scattered pieces. Going on for the next 4 years, scattered here and there throughout Greece, the outbreak continued to hamper the daily lives of the citizens. Originated in sub-Saharan Africa, south of Ethiopia, the disease found its way to Greece, Persia, and the Mediterranean region by spreading through the north and west of Egypt and Libya. During this period, Athens witnessed an epidemic with a wiping out of nearly one-quarter of its population, around 75,000 to 100,000 people, because of the plague. Around 30 pathogens have been found to be responsible for the outbreak. The plague not only killed the masses, but it also affected the equation between Sparta and Athens. The greatest price was paid by the leader of Sparta, Pericles, who became a victim to the infection at an early stage, pushing the fate of Sparta into the abyss. Affecting the societal norms of Athens, the plague caused a divide between strict laws and their adherence. In order to protect the inner walls of Athens, Sparta was a land-based army. Relying on Athenian maritime supremacy, and focusing solely on the land, causing overpopulation, close quarters, unhygienic situation, shortage of food supplies, cost Athens its culture, civilization, and much more. Becoming a heavy breeding ground for the epidemic became inevitable. People ceased to fear the law and religion as they felt that due to the plague they were already under a death sentence. Temples represented a sight of abundant misery and suffering, filled with the dead and dying. In these moments of rising superstitious activities and despair, the balance of power shifted, resulting in Athen's loss of power and capabilities to expand its territorial powers. A number of people succumbed to slavery, there was a loss of morale among the army, and an exponential increase in corruption gripped the city that was once an epitome of cultural and intellectual activities. The plague did not stop causing havoc just here. It made a return twice in the land of Athens, in 429 BCE and 427 BCE. Without proper planning and lessons learnt, the devastation was equal in comparison to the first outbreak. Its accounts have been mentioned by one of the survivors of the diseases; Greek historian Thucydides recorded his experiences in a tremendous work named *History of the Peloponnesian War*. He narrated the horrifying scenes with much information on his hands, having suffered from the plague himself. The symptoms of the disease mentioned by the historian in his book paint a ghastly picture of the horrifying condition the patients were in. The patients suffered from a never satiating thirst that no amount of liquid consumed could fulfill. A continued restlessness gripped the sick followed by insomnia, a great difficulty to sleep soundly. In this situation, it was not a surprise that the sick died after 7–9 days of catching the disease, having to suffer endlessly throughout the period they had the symptoms. A violent ulcer and severe diarrhea followed those who were fortunate enough to survive, but not for so long. If not the heat and fever, ulcer and diarrhea caught up with the patient eventually and their death became imminent. Even if someone was fortunate to survive the outbreak, they were left with disfigured bodies, blindness, and memory loss. The birds and animals who consumed the diseased flesh also caught the plague and could not live long enough.

This brought the historians and researchers to question the kinds of diseases Athens was swept in. Was it a single disease that destroyed a whole culture or was it the combination of diseases working together to replicate faster and harder? As Thucydides was not professionally trained in medicine, his description of the symptoms is based on his own experiences and what the patients narrated to him. In the light of this information, it becomes evident that he has not noted down how the physicians tried to overcome the disease with various experiments but failed in the process—many of them succumbing to death themselves as the plague engulfed anyone and everyone who were coming in close contact with the sick, turning the plague into a contagious and deadly outbreak. Although, he did mention that the disease was of a kind that the physicians had never encountered before. He recorded his account in order to gain insights for the future generation if something similar of a situation emerged again. The war itself did not help. Both the armies, wanting each other to die at the earliest, started poisoning the wells. This led to the death of even healthy people.

The devastation caused by the epidemic had been recently corroborated with an excavation of mass graves in Athens in 1994–95. As it was one of the first epidemics witnessed by the physicians, they had no clue what procedure to follow and how to treat the diseases. As they were in close contact with the victims, many of them contracted the disease and died. It took around 15 years to completely recover from the devastation. Historians believe that by far, this was the most devastating and lethal period of illness Greece had ever suffered from Ref. [6].

Typhus, smallpox, measles, toxic shock syndrome, fungal infection, and even Ebola have been linked to this epidemic. In "The Plagues of Athens," J.F.D. Shrewsbury, eliminates smallpox as a leading cause, because the patients of smallpox do not have the strength to get up from the bed. Submerging in the cold water is also not linked with smallpox, also backache has not been mentioned which is a main symptom of smallpox. He eliminated typhus because personal hygiene was not a problem in Athens, nor is there any evidence of excessive lice growing in the black rat population. Deafness is more of a symptom of typhus, contrary to the blindness described in the book. Bubonic plague was similarly eliminated for the lack of evidence of the presence of black rats along with pneumonia, as coughing and spitting of blood was not mentioned. Typhoid was eliminated because unclean waterways were not described. He settled on measles which has the similar symptoms accounted for by Thucydides. Some researchers also argued that the description does not match any modern diseases. An evolution in the working of the host and parasite makes it difficult to identify and pinpoint any particular disease as a cause of an outbreak. A high mortality rate is associated with smallpox, occurring with fever and rash; however, the patients develop immunity afterward. Moreover, humans are only known to be the host for smallpox, but going by Thucydides' narration, birds and animals seem to be affected too. He has mentioned a mortality rate of 25% among the soldiers but there is no data on children, who are most likely to succumb to the disease. Bubonic plague affects both humans and animals, but fleas are the main carrier of the disease rather than a human-to-human transmission. Measles occurs in densely populated cities [5]. The reason to rule out typhus was that it causes rash and red spots, but Thucydides mentions rashes and blisters. Ergotism is not contagious and the survivor does not develop immunity, therefore it was eliminated too.

As all these diseases individually have been ruled out, a possibility of existence of multiple diseases simultaneously grows strong. It is not a rarity for several diseases to exist at the same time, and developing an immunity for one disease does not mean it will secure a patient from another disease. There is also a strong possibility that the previous microbe has changed its nature along with its symptoms in these 24 centuries or has gone extinct. If researchers try to compare the earlier and modern microbes, it is likely that they will see more dissimilarity than a similarity. It should be kept in mind that it is extremely difficult to identify a disease at the early stage as many diseases have a common symptom. Thucydides' information seems to be biased as he himself suffered from the illness and narrated the situation with a literary perspective. His description does not apply to one but multiple diseases. One disease can clear the ground for other diseases to occur, so going after only one disease will be amateurish.

To date it remains a mystery as to what actually caused the plague in Athens. It should be noted that all the diseases held responsible have not been diagnosed as lethal. The reliability of Thucydides accounts also come into question because of his nonmedical background, making it a nearly 2500-year-old controversy. An absence of an RNA test on the viruses present during the plague reduces the probability of finding what conspired in Athens even further.

Second plague pandemic

Haunting European, North African, and West Asian countries throughout the 14th and 17th century, the plague kept returning via ports, making it responsible for wiping out two-thirds of the population [7].

Black Death

Centuries later, the same bacterium, *Y. pestis* revived itself to cause another bubonic plague, in Afro-Eurasia, from 1346 to 1353, killing 75–200 million people, the most fatal of the epidemics, painting this part of human history a shade grimmer [8]. Affecting the historical, socioeconomic, and cultural course of Europe, the pandemic most famously known as Black Death, killed a significant number of people in over half a decade. The plague is known to have spread through fleas on black rats living in slave ships, traveling around the Mediterranean ports. Once reaching the land, the plague was spread by fleas that turned the plague into a pneumonic plague wiping out nearly 40%—60% of the European population; the continent took another two centuries to recover its previous population digits [9].

Y. pestis during Black Death is ancestor to all the current circulating strains of *Y. pestis*. The existing plagues can find an origin in the medieval period. There were multiple recurrences of the outbreak throughout the 14th century. The fluctuation in the climate intensifying the population of rodents and plague-carrying fleas also did not help the situation. Mortality rates vary from place to place depending on the extent of isolation the city was in. The monastic communities proved to be the breeding ground for the plague as they were densely populated. The plague affected rich and poor alike. Even if the royalty was privileged enough to leave the town, they could not. Annihilation of the entire family was a common scenario. Diaries filled with the names of people who died during that period reveal the ravage the infection created in the European land.

Lymphatic nodes are the most affected in a bubonic plague. In case it is not treated, it affects the lungs and blood severely. Blood and pus seeping out of the bodies were followed by fever, chills, vomiting, diarrhea, terrible aches and pains, and eventually death. So was the devastation caused by the plague that healthy people going to bed at night used to be dead by the morning.

The Italian poet Giovanni Boccaccio described the horrifyingly and efficiently contagiousness of the infection—"The mere touching of the clothes appeared to itself communicate the malady to the toucher." Not knowing how to treat the disease, even the physicians succumbed to superstitions and mysticism, narrating the spread of the terrifying disease resulting in death as, "instantaneous death occurs when the aerial spirit escaping from the eyes of the sick man strikes the healthy person standing near and looking at the sick." Their treatments were inspired by superstitious beliefs such as burning aromatic herbs and bathing in rosewater. Soon the local economies failed when the doctors refused to see patients, priests closing their doors, and shoppers shutting down their businesses. Fleeing to the isolated countryside did not prevent the vast number of deaths. The disease did not discriminate between humans and domesticated animals, affecting them the same. The wool industry plummeted.

Reaching Rome and Florence, which were the most active and center of the trading routes, the infection knew no bounds. It spread like a wildfire while people were still contemplating and making sense of everything that was happening around them. One of the major economic changes that the society encountered was a reduction in the number of laborers. This became the cause of ruin for many landowners. The wages of artisans and peasants increased because of high demand. The lines between the middle class and poor blurred and the rigidity in the stratification of the society started to vanish. In poetry, sculptures, and paintings, the artists started retelling their experiences in the face of the plague. It led the Roman Catholic Church to lose its power over people as they turned to mysticism. Some people scourged themselves after believing that the plague was a result of humanity's sins—a punishment from God. They moved from one city to another to display their penance by beating themselves with a leather strap filled with sharp metals while the public watched them. The Jew community was blamed for spreading the disease. Many of them were killed on the spot, their entire communities burned to death. Fleeing to the eastern Europe was the only option left for them. Every night a crier would roam around the town and call out the family members to "bring out the dead" for mass burial.

The recurrence of the pandemic caused economic havoc in England in the mid-15th century. In the country alone, the plague reduced the population or total disappearance of nearly 1000 villages. It is difficult to accurately predict the number of deaths but an estimate is around 25 million [10]. The disease could not be stopped completely but it was slowed down by isolating the arriving sailors and checking up on them for any illness for days. Social distancing and isolation assisted in reducing the spread widely. Even though there exists proper treatment for the disease, it has still not been eliminated entirely. Every year, around 1000–2000 cases of bubonic plague appear according to the World Health Organization [11].

Third plague pandemic

Between 1855 and 1859, the pandemic originated in China and spread across all the continents with human inhabitants. Spread during the Qing Dynasty, the pandemic was responsible for nearly 12 million deaths in China and India, majority of which were in India. The WHO claimed that the pandemic was active till the 1960s after which cases dropped to 200 per year [12]. There can be two sources of this plague: the bubonic plague spread through ocean-trades which transported infected sailors, black rats, and fleas to and fro. The pneumonic plague was more contagious in nature but confined within Asian countries—Mongolia and Manchuria. Yunnan region in China was densely populated to exploit the minerals in the 19th century. Around 7 million people inhabited the area while transportation increased exponentially. When infected people reached isolated cities, the disease killed the entire cities. Within 2 months, it had engulfed other regions of China. There were nearly 100, thousand deaths when the epidemic stopped but Hong Kong witnessed the pandemic until 1929.

It was after half a decade since the beginning of the third pandemic that French-Swiss bacteriologist Alexandre Yersin and team in Hong Kong researched about the pathogen. San Francisco witnessed its first pandemic between 1900 and 1904 and another between 1907 and 1908. In 1898, French researcher Paul-Louis Simond demonstrated how fleas worked as vectors for the bacteria. The social implication of the plague was that it increased the racial bias between the colonizers and citizens.

Due to the fear of the pathogens developing drug-resistance against antibiotics, its heavy usage was discouraged. In 1995, such a drug-resistant bacteria was found in Madagascar. The first influenza epidemic, in Europe, between 1556 and 1560, caused a great deal of death.

Smallpox

Caused by the variola virus, smallpox was a terrifyingly contagious disease in the 18th century spreading from person to person with a fever and skin rash. Although an accurate origin of smallpox is debatable, medical writings from India and China around 1500–1100 BCE shed light on the possible origin of the disease. It could also have existed more than 3000 years ago because of the evidence found in an Egyptian mummy of Ramses V. Egyptian traders brought smallpox to India is one of the speculations. At the same time, China was also being swallowed by the infection, taking it to Japan as well, where the infection wiped out one-third of the population. It would go on to become an infectious disease in the Indian subcontinent for at least 2000 years. So much was its terror that people started worshiping as many as seven deities to cure them of this terrible disease.

In the 18th century, smallpox killed around 60 million people, leaving an estimated one-third of the survivors blind. It became a deadly disease in 18th century Europe killing 400,000 people each year, while 10% of the Swedish infants succumbed to death and death rate in infants in Russia rose higher than the birth rate [13]. Affluent families in North America, Great Britain, and China were among the first to get treated with variolation. By the end of the 19th century, when vaccination finally became common in most parts of the world, the infection started to die down. A taint in the history is that smallpox had been used as a biological weapon at various wars where soldiers were ordered, paid, and sanctioned to use smallpox against the enemy.

In China, the practice of variolation was happening during the 10th century. Historical texts inform about powdered smallpox swabs being blown up to the nose of healthy people, so that they can develop an immunity for it. This method of herd immunity reduced the mortality rate of 20%–50% to about 0.5%–2.0%. Such practices were also introduced in Turkey and the UK. Efforts toward vaccination have helped in reduction and eradication of the persistence of the disease. It only remained in the Horn of Africa by the end of 1975 and in 1979, after rigorous activities in verification of existence of any cases, a global eradication of smallpox was certified and later endorsed by the World Health Assembly.

At an early rash stage, the patient poses a heavy risk of spreading the contamination to others. Prolonged face-to-face contact results in direct transmission of the disease. Once the patient has sores in their mouth and throat, they become contagious. Coughing, sneezing, and contaminated droplets from mouth and nose are responsible for spreading the virus. Until and unless the last scabs of smallpox get removed from their body, the virus can find an opportunity to develop and thrive. The scabs and the fluid in them contain variola virus that could spread from smallpox patients to the physicians caring for them. In order to be safe from the infection, the medical staff had to wear gloves, masks, and change their clothing and beddings multiple times a day. The virus could also spread through air in an enclosed area such as a building. This is one of the diseases that can spread from humans to humans only. Animals and insects are not known to transmit the disease.

The smallpox vaccine was made from another virus called vaccinia, a poxvirus with little harm. However, vaccination does not mean that someone is completely safe from the disease. Vaccinated people have to take proper precaution when caring for the smallpox patients. Although there were rarely any life-threatening adverse effects of the vaccination, a few vaccinated people developed heart inflammation (myocarditis), inflammation of the lining of the heart (pericarditis), or a combination of both (myopericarditis).

Tuberculosis

The 19th century was scarred by tuberculosis, caused by lung affecting bacteria, *Mycobacterium tuberculosis*, which wiped out one-quarter of the European population, proving it to be one of the permanent stains in the course of human history. The origin remains unclear due to the bacteria's long-term existence. German physician and microbiologist, Robert Koch discovered the *M. tuberculosis* bacteria in 1882 which got him a Nobel Prize in Physiology or Medicine in 1905. Many of the famous poets and writers called the disease "the romantic disease" as they either had caught TB or knew someone who had [14].

Spreading directly from person to person when the infected person sneezed, coughed, spitted, or transmitted the infected droplets to a healthy person, the disease caused havoc in a short period of time. A few of the germs can infect a person gravely. About a quarter of the world population is infected with TB bacteria but they would hardly show any symptoms. Around 5%—10% of the population has a risk of falling ill and a higher risk among immunocompromised patients. A delay in seeking medical help can increase the chances of spreading the disease to others. The rate at which the disease spreads has associated many stigma in societies in developing countries. In some countries, TB patients are not allowed to attend public gatherings. Such stigmas and long-term treatment have prevented TB patients from seeking medical assistance at an early age and the relatives keeping the cause of death a secret so as not to face social ostracism.

Developing countries are at a higher risk with over 95% cases and death occurring there. Similarly, the amount of risk becomes even higher in HIV patients or any disease that prevents the immune system from functioning properly. A lifestyle with usage of alcohol and tobacco increases the risk factor. Coughing with sputum and blood at times, chest pains, weakness, weight loss, fever, and night sweats are some of its symptoms. In 2019, tuberculosis killed 1.4 million people among the 10 million total patients. TB has not discriminated between different age groups and continents. In the same year, India had the most number of TB patients. Multidrug-resistant TB, a severe kind of TB, is a public health crisis. Proper sanitation, vaccination, and other public health measures can reduce the spread of TB even before the arrival of antibiotics.

TB is curable and preventable. With TB diagnosis and treatment, between 2000 and 2019, around 60 million lives have been saved. One of the ambitious targets for United Nations Sustainable Development Goals is to eradicate TB by 2030 [1].

During this era of early 20th century a lot of treatment options also came into light. With the discovery of penicillin the entire focus shifted toward antibiotic production and antibiotic discovery. In comparison to bacterial drug discovery antifungal drug discovery has always suffered a lacuna. But after the middle of 20th century there was a spike in cases of antifungal infections which led to scientific focus of these infections under the microscope. Subsequently lot of fungal infections were identified, characterized, and classified. A lot of antifungal drugs and their classes were also discovered.

Spanish flu

The most infamous, Spanish flu or the influenza pandemic of 1918 killed around 50 to 100 million humans, wiping out around 2% of the world population of 1.7 billion. Caused by the H1N1 influenza A in four successive waves, the virus infected 500 million people in a span of only 2 years around the world which was one-third of the human population of Earth. With no effective treatment in sight, the citizens were asked to wear masks, follow social distancing, and shut up schools, theaters, and businesses. While the number of Spanish Flu deaths reduced 1 year after its occurrence, the Justinianic plague lasted for half a century.

Flu virus is highly contagious. It can spread through droplets in the air when the infected person sneezes, coughs, or talks. Touching the face after being in contact with an infected surface can also infect a person. Immunocompromised patients of asthma, tuberculosis, pneumonia, sinus infection suffer from a flu infection greatly.

At first, the number of death cases was low even though the flu was spreading relatively faster infecting a majority of people. The patients had fever, chills, and fatigue and they recovered after several days of infection. This was not going to be the case for long because soon a more deadly form of influenza returned to haunt the world, killing the patients after hours and days of displaying symptoms. Their lungs were filled with fluid that caused discomfort during breathing, their bodies turning blue and suffocated the life out of people, reducing the life expectancy of people by a considerable number.

The pandemic was worsened by the censoring of the reports for keeping the morale high during World War I. In Spain, a neutral tone was adopted by the newspapers to report on this deadly disease, which is why the pandemic came to be known as Spanish flu. It has been observed that the influenza infects the immunocompromised or the very young, or the old, because they have a weaker immunity to combat deadly pathogens. But Spanish flu went beyond one's expectations and former knowledge. It affected the young adults the most. Researchers believe that it might be because of a climate anomaly lasting for 6 years or due to an increase in waterborne diseases. Trigger of a cytokine storm in the immune system has also been a possible cause. Malnourishment, overcrowded medical camps and hospitals, and poor hygiene aggravated the already burning situation.

Young people who were previously thought to be immune to influenza were hit the worst. The disease killed more people than World War I in which nearly 16 million people died. A lack of sophistication in medical record-keeping lowers our chance to accurately know the exact number of fatalities around the world. The origin of the pandemic is still debated with researchers having theories of its origin in France, Britain, China, and the United States. The World War I soldiers were most likely to spread the disease by going from one camp to another. When the flu first made its appearance, the doctors had no idea of the cause or how to treat it. It was not until the 1940s when the first licensed flu vaccine came in the market. The shortage of physicians was so much that the medical students were appointed as staff, and schools were converted into temporary hospitals. Public places were shut down and citizens were forbidden to gather in masses. "Sanitary Codes" were implemented by Boy Scouts.

Bridging the lack of a vaccine, doctors prescribed aspirin to reduce the pain. Patients were recommended to take 30 g of dose a day, which has been proven to be deadly now. Any prescription of more than 4 doses can be fatal according to the present medical consensus. Aspirin turned out to be more harmful for the patients as it built up more fluid in lungs, causing hyperventilation and pulmonary edema. If not caused, at least aspirin poisoning accelerated many deaths.

The Spanish flu witnessed humanity at its lowest. With the number of family members dying every day, the burial grounds faced a shortage. Many times, an entire family would be wiped out and other times, people had to dig graves for their own family members. With corpses piling up in the graveyard, the economy took a toll too. With sanitation workers falling sick, the basic activities of sanitation were difficult to perform. Farmers were not able to harvest crops. Businesses were closing down faster than ever, which prevented proper documentation on the pandemic. Downplaying the disease at the beginning in order to not create a public outcry turned out to be fatal. Fines as high as $5 were charged to the rule breakers in San Francisco.

The horrific time came to an end in the summer of 1919 after people developed immunity. There have been a number of influenza pandemics, but not as devastating as the Spanish flu. Going on for a year, a pandemic in 1957 killed 2 million people globally. One decade later, another pandemic in 1968 was responsible for the death of 1 million people.

H1N1 Influenza A virus went on to cause another of two flu pandemics. The Russian flu in 1977 affected young children, and the most recently, 2009 swine flu pandemic.

HIV/AIDS

HIV (Human Immunodeficiency Virus) compromises the ability for one's immune system to fight back, which leaves the patient vulnerable to other diseases. A point of no return is when HIV completely destroys the immune system, reaching to the last stage of infection called AIDS. Among all the infectious diseases humans have suffered with, HIV is one of the recent. Never in the history of infectious diseases was it easier for a disease to spread than it was in the wake of the 21st century. A changing lifestyle of air travel and narcotics dependency did not help the situation. But the brunt was faced and the heavy prices were paid by the developing nations with the poorest of the communities; people who suffered from the disease experienced a drop of life expectancy by 20 years. There exists no effective cure for HIV/AIDS but early detection and antiretroviral therapy can help people deal with this disease better and prevent transmitting it to their sexual partners. Taking HIV medication also reduces the chances of reaching the last stage of the disease. Without medication, patients typically survive about 3 years and in case of an arrival of an opportunistic disease, the life expectancy reduces to 1 year. To know if someone has HIV or not is by getting a blood test done. Self-test kits are also available in the market. Fever, chills, night sweats, muscle aches, sore throat, fatigue, swollen lymph node, and mouth ulcer are some of the symptoms with a habit of making an appearance and disappearing within 2—4 weeks.

At first, a gay magazine published about a rare disease with only five cases. Soon after, an unexpected number of homosexual men were being admitted to hospitals. Earlier a term called GRID which stands for gay-related immunodeficiency was coined for this mysterious disease. But researchers later named it AIDS—Acquired Immune Deficiency Syndrome. According to researchers, HIV was transmitted from a chimpanzee to a human in West Africa through hunting in the 1930s. The virus found a breeding ground in the 1980s when strange cases of pneumonia, cancer, along with other illnesses gripped the patients. The illness knew no mercy. The first case was reported in the United States in 1981 and first thought to be transmitted only in gay men. Next year, the cases increased and researchers began to understand that the virus was being transmitted sexually, through needles used for injecting drug, and can be transmitted from mother to child through breastfeeding. Starting as a zoonotic disease, when the microorganism starts adapting to humans as a host, the situation was grave. A retroviral drug called AZT was approved for use by the FDA. Just before the beginning of the 21st century, the WHO had declared HIV as the fourth biggest cause of death and number one in Africa, having killed an estimated 25 million people between 1982 and 2002 according to UNAIDS while an estimated 42 million people lived their lives under the shadow of a grave disease in 2002. In 2016, worldwide the number of people suffering from HIV was an estimated 36 million. To prevent and reverse its spread, the disease was adopted by the United Nation in its Millennium Development Goals along with TB and malaria. Negotiations to lower the cost of drugs were made for the developing countries. In 2014, 90—90—90 targets were launched by UNAIDS, aiming to diagnose 90% of people living with HIV, to provide accessible medical treatment to 90% of those patients and to achieve viral suppression for 90% of patients getting medical treatment by 2020. Antiretroviral treatments have also proven to have reduced the gravity of the situation.

The societal and medical stigma related to HIV has been profoundly grave. From shunning, rejection, discrimination, to mandatory HIV test without pior consent, violence against HIV infected patients and quarantining, the patients were only introduced to a life of despair and vulnerability. The disease was worsened by this stigma when such discrimination prevented the patients from seeking medical assistance. Homosexuality, bisexuality, prostitution, and intravenous drug use were met with severe criticism and abhorrence. The economic impact for individuals and countries has been immense. A reduction in global GDP, lack in human capital, and increase in unemployment rate among the patients meant a considerable economic loss for developing nations. Whereas, individuals faced reduction in self-esteem, sense of confidence, self-dignity, and quality life. The number of HIV infected orphans is frightening.

Swine flu

In a span of 19 months, swine flu killed more than 18,449 people who were lab-confirmed by the WHO while the estimated number of total deaths is more than 0.24 million. It was first reported in the United States. Due to a reassortment of human flu, European pig flu, and bird flu, a new strain got developed. It is estimated that including the asymptotic and mild cases, the influenza has affected anywhere between 700 million to 1.4 billion people, around 11% of the population of 6.8 billion people. The most unusual observation about this pandemic was that it was not targeted at the very old people. Healthy young people who were infected with the virus also developed pneumonia.

Named for the composition of the proteins hemagglutinin (H) and neuraminidase (N) that form the viral coat, the virus is a subtype of influenza. Since the 1930s, four types of flu viruses have been isolated—H1N1, H1N2, H3N1, and H3N2. The symptoms of the flu include coughing, fever, nasal discharge, and illness lasting for about a week. The flu spreads rapidly among pigs and whoever comes in contact with the pig, contaminated food, or bedding or inhales the infected droplets in the air gets infected. The flu is known to cause diarrhea, vomiting, and chills as well. There is no particular treatment for swine flu. Isolating the infected pigs from the healthy pigs and administering antibiotics can prevent the flu from further spreading.

COVID-19 (current pandemic)

A novel coronavirus, severe acute respiratory syndrome-coronavirus-2, started a pneumonia outbreak in China in December 2019 that quickly expanded around the world. Asymptomatic instances or moderate symptoms such as fever, cough, sore throat, headache, and nasal congestion to severe cases such as pneumonia, respiratory failure requiring ventilator support, multiorgan failure, sepsis, and mortality are among the clinical characteristics of the disease. Because the rate of transmission is so high, we need an efficient treatment strategy for treating symptomatic individuals as well as preventative measures to contain the illness and stop the spread [15]. This unique coronavirus disease, also known as coronavirus disease 2019 (COVID-19), has spread rapidly over the world due to its high transmissibility. In terms of both the number of sick people and the geographic span of epidemic locations, it has overwhelmingly surpassed SARS and MERS [16]. In the 21st century, the current COVID-19 epidemic is destined to become a key anchor point. It serves as a reminder of how swiftly diseases can spread over the world as they interact with economic, political, and cultural forces. It also reveals significant differences in people's perceptions of and responses to a new health threat, as well as the deep politicization of responses at the local, national, and global geopolitical levels. Historical comparisons raise questions about why and how human, communal, scientific, and societal responses to such threats differ over time [17]. Pandemics have always had an unequal impact, both now and in the past. Some pandemics, such as Ebola and MERS, wreaked havoc in certain areas or a small number of nations before being confined and spreading to other parts of the globe. Pandemic containment in the 20th and 21st centuries was aided by past experiences with unchecked spread. Infectious disease epidemics have shaped our communities and cultures like few other events in human history; yet, surprisingly little attention has been paid to these events in behavioral social science and fields of medicine that are, at least in part, established in social studies [18].

A note on the future

When we go through the early history of infectious disease we find the beginning of era of plague and uncontrollable dissemination. The past is a great instructor, even if we do not want it to be. There has been a high improvement in public health and hygiene since the medieval times. Because the bacteria was now an extinct form of *Y. pestis* and even though it still is around, it is highly unlikely that a similar kind of devastation would return in the future. After the world was hit by coronavirus, the interest in epidemiology is increasing. A history of epidemics that had been overlooked in the historical texts for so long is making its way again in hopes to raise awareness about public health and individual responsibility toward hygiene. Taking lessons from history, each one of us can make educated and more efficient decisions, learning each time not to repeat the mistakes made by our ancestors. One of the lessons to be learnt is that the pandemics are not discriminatory. Their effect, severity, and burden are equally shared by the rich, poor, urban, rural, socialist, or capitalist communities. Unless a herd immunity is developed and a majority of people are vaccinated, pandemics would keep making comebacks.

Every day our fight with the microorganism world is taking a different shape, a different form and why shouldn't it be? They have been existing on Earth for thousands of years more than we do, making a world of their own, replicating every time they get a chance and shrinking the population of animals and humans. Their much longer existence on the planet should only remind us of their capabilities in adapting to changes through and through. Both of these worlds' constant struggle to survive and overpower each other will have a fate based on our technical improvement and never giving up on our existing knowledge and a will to fight, because the pathogens certainly are not giving up. Their evolution from one strain to another, possessing more strength and immunity toward our own immune system is evidence to that. Hibernating for several decades, even centuries, these pathogens have a history of returning with much force and fatality which is why we should strive to be better prepared for any such upcoming incidents.

COVID-19 has shaken up the world, taking up by the storm, devastatingly surprising the common people and causing fatalities and social disorder of a scale that the world witnessed nearly a century ago. With next to zero recollection of the past, handling of the situation in various countries has been poor. A lack of knowledge, preparation in the health sector, and willingness in people to follow the quarantine guidelines seem to be one of the major reasons the world witnessed its terrifying effects.

The pathogens live, survive, and thrive near us every day in rodents, bats, swines, and multiple other animals. The world is more of microorganisms' than ours, yet we choose to ignore them so blindly that they have to create a pandemic for us to notice, learn, and improve our game or face death. It would only be a foolish mistake to be not prepared for more of such pandemics. The horrific amount of devastation and havoc these pandemics have created and still create should only intensify our efforts and advocacy for research and development.

In a sample of 1100 bats from Yunnan province in China, were found a number of coronaviruses, out of which four were closely related to the coronavirus that caused the ongoing pandemic. If the whole of Southeast Asia's bats would be tested for such viruses, one cannot imagine what Pandora's box they would reveal. It is just the scratch of a surface we have been exposed to now. The probability of transmission of viruses from animals to humans is pretty high. We never know which one of such transmissions can hamper our worlds with devastating consequences. Even if the transmission of pathogens does not cause any pandemic, the humans are still becoming a guinea pig and a breeding ground for the pathogens to adapt and survive in a new and changing environment of the host; and analyzing their survival history, the picture does not look any good. From then on, finding a dense city for the pathogen is the easiest of the tasks. With new variants coming up in an interval of several months, it is evident that the research in this field must be our first priority. We need to gather ourselves up and be prepared for any kind of pandemic hitting. More funding into research, encouraging collaborations, awareness among people, and public hygiene are some answers to be better prepared for the pandemics.

References

[1] L. Shaw-Taylor, An introduction to the history of infectious diseases, epidemics and the early phases of the long-run decline in mortality, The Economic History Review 73 (2020) E1, https://doi.org/10.1111/ehr.13019.

[2] J. Piret, G. Boivin, Pandemics throughout history, Frontiers in Microbiology 11 (2021) 3594, https://doi.org/10.3389/fmicb.2020.631736.

[3] I. Ansari, G. Grier, M. Byers, Deliberate release: plague—a review, Journal of Biosafety and Biosecurity 2 (2020) 10−22, https://doi.org/10.1016/j.jobb.2020.02.001.

[4] R.D. Pechous, V. Sivaraman, N.M. Stasulli, W.E. Goldman, Pneumonic plague: the Darker side of *Yersinia pestis*, Trends in Microbiology 24 (2016) 190−197, https://doi.org/10.1016/j.tim.2015.11.008.

[5] J. Piret, G. Boivin, Pandemics throughout history, Frontiers in Microbiology 11 (2021), https://doi.org/10.3389/fmicb.2020.631736, 631736.

[6] D. Wright, Infection control throughout history, The Lancet Infectious Diseases 14 (2014) 280, https://doi.org/10.1016/S1473-3099(14)70726-1.

[7] S.N. DeWitte, Mortality risk and survival in the aftermath of the medieval black death, Plos One 9 (2014) e96513, https://doi.org/10.1371/journal.pone.0096513.

[8] C.J. Duncan, S. Scott, What caused the black death? Postgraduate Medical Journal 81 (2005) 315−320, https://doi.org/10.1136/pgmj.2004.024075.

[9] B.P. Zietz, H. Dunkelberg, The history of the plague and the research on the causative agent *Yersinia pestis*, International Journal of Hygiene and Environmental Health 207 (2004) 165−178, https://doi.org/10.1078/1438-4639-00259.

[10] K.A. Glatter, P. Finkelman, History of the plague: an ancient pandemic for the age of COVID-19, The American Journal of Medicine 134 (2021) 176−181, https://doi.org/10.1016/j.amjmed.2020.08.019.

[11] D.J.D. Earn, J. Ma, H. Poinar, J. Dushoff, B.M. Bolker, Acceleration of plague outbreaks in the second pandemic, PNAS 117 (2020) 27703−27711, https://doi.org/10.1073/pnas.2004904117.

[12] B. Bramanti, K.R. Dean, L. Walløe, N. Chr. Stenseth, The third plague pandemic in Europe, Proceedings of the Royal Society B: Biological Sciences 286 (2019), https://doi.org/10.1098/rspb.2018.2429, 20182429.

[13] R.A. Weiss, J. Esparza, The prevention and eradication of smallpox: a commentary on Sloane (1755) 'An account of inoculation, Philosophical Transactions of the Royal Society B: Biological Sciences 370 (2015), https://doi.org/10.1098/rstb.2014.0378, 20140378.

[14] T.M. Daniel, The history of tuberculosis, Respiratory Medicine 100 (2006) 1862–1870, https://doi.org/10.1016/j.rmed.2006.08.006.

[15] V.M. Mahalmani, D. Mahendru, A. Semwal, S. Kaur, H. Kaur, P. Sarma, A. Prakash, B. Medhi, COVID-19 pandemic: a review based on current evidence, Indian Journal of Pharmacology 52 (2020) 117–129, https://doi.org/10.4103/ijp.IJP_310_20.

[16] B. Hu, H. Guo, P. Zhou, Z.-L. Shi, Characteristics of SARS-CoV-2 and COVID-19, Nature Reviews Microbiology 19 (2021) 141–154, https://doi.org/10.1038/s41579-020-00459-7.

[17] E. Frankema, H. Tworek, Pandemics that changed the world: historical reflections on COVID-19, Journal of Global History 15 (2020) 333–335, https://doi.org/10.1017/S1740022820000339.

[18] D. Huremović, Brief history of pandemics (pandemics throughout history), Psychiatry of Pandemics (2019) 7–35, https://doi.org/10.1007/978-3-030-15346-5_2.

CHAPTER 2

Brief introduction on infection

Introduction

The principles of infection control are based on significant research and should be followed to effectively prevent and control infections. Infection management is especially crucial in hospital settings, as patients are at a higher risk of infection. Understanding the infection process should lead to the implementation of suitable measures to safeguard patients and healthcare personnel. Infection control strategies used during patient care can go a long way toward preventing or eliminating nosocomial infections.

Infection prevention and control strategies are critical in the daily care of patients in healthcare settings. Many diverse microbes constantly share our environment with us. It is critical for healthcare personnel to understand them and their pathogenicity. Awareness of the infection cycle, as well as the use of proper precautions and disinfection methods, is critical. Infection control can be influenced by the quality and quantity of accessible resources and facilities. However, every precaution should be taken to prevent infection. Basic infection control techniques are the best defense against pathogens that are extremely resilient. The overall health and nutritional status of the public are essential in prevention, containment, and infection control. Such factors determine the prevalence of infectious disease in the population, which has an impact on the infection rate of people both in and out of hospitals, putting a strain on healthcare institutions.

Occurrences of the past 30 years show that our infectious disease defenses cannot be lowered. Infectious disease control and prevention are improving, but we must not get complacent [1]. Researches all over the world have vividly demonstrated infectious disease's ongoing impact on the world. These studies have helped

CONTENTS

Introduction ... 27
References 31

Anticandidal Therapeutics. https://doi.org/10.1016/B978-0-443-18744-5.00011-3
Copyright © 2023 Elsevier Inc. All rights reserved.

in understanding how new studies can aid in the development of new control and preventative strategies. As researchers, we take risks in our fight against infectious diseases. Smallpox has been eradicated, and two other diseases are on the verge of extinction [2]. However, we must set realistic goals, which mean that the start of an eradication program should be limited to the few diseases for which this is a viable goal. As we plan our continuous commitment in our approach to infectious diseases, we should place a strong emphasis on control and prevention.

Gradually epidemiological studies were reported and based on those studies, observation, monitoring, and care of patients was slowly incorporated in the preventive and treatment measures [3]. Initially a lot of emphasis was given to the epidemiology of bacterial diseases [4]. In the early ages the only infection that was identified was caused due to bacteria which include smallpox and salmonella. Eventually, as time progressed, development of statistics and surveillance helped in diagnosis of diseases in early stages. This helped in prediction of a disease and thus better management.

As we studied in the last chapter, during this period many new discoveries were being made. The first report on infectious microorganism was made by Anton von Leeuwenhoek in the year 1683. He first discovered microorganisms like bacteria. Later on with the likes of scientists like Alexander yersin [5], Louis Pasteur, and Shibasaburo Kitasato [6], infectious organisms were discovered. In the future decades many scientists began to focus specifically on the vector-borne diseases; hence a golden age was born where a lot of bacterial and viral diseases were discovered along with their infectious organism and their host [7]. These diseases include Texas cattle fever [8], yellow fever [9], lyme disease [10] cat scratch fever, human granulocytic and monocytic ehrlichiosis [11].

In the 20th century with the development of National institute of health and several disciplines of microbiology, virology, and immunology the emphasis on the precise nature of infection came to light. With this advancement at that age, clear bifurcation of viral, bacterial, and fungal diseases was made with an aim to emphasize each nature of infection.

Now the challenge that light lies ahead of us is to not only treat the current infections but also to eliminate future threats. The biggest challenge that still remains is antimicrobial resistance. Pathogens have been evolving with time and due to several underlying reasons they develop partial or complete resistance against current treatment regimen. This makes it next to impossible to completely eliminate such threats. Scientists all over the world have been constantly making efforts to solve the case of antifungal resistance permanently. For this complete understanding of epidemiology and molecular biology and pathogenesis is important. In this book authors are trying to give a detailed account on one specific fungal infection causing pathogen which will help future researchers to understand every aspect of its infection.

Even in the earliest times when modern facilities and resources were not available human beings have sought to understand the natural forces and risk factors responsible for diseases and infections. This gradual understanding has led to discovery of antibiotics and relevant drugs and also the identification of a disease and its causative organism. This basic information alone is the first step of its cure, although advances in knowledge and advancement in resources have always hindered this process. In limited resources the emphasis of research has always been toward conventional bacterial infections and fungal-based infections have always been slightly ignored. But in those early times measles and smallpox have been affecting the world as the most devastating epidemic diseases. Hence the view of the world also gradually and automatically shifted toward treating bacterial infections. Invention of microscope created a scientific revolution where visualization of microorganism became possible; for the first time it brought upon immense knowledge that helped in cultivation identification of microorganisms.

In 21st century we are facing a problem like no other. Emergence and reemergence of infections have been going back to back in the circle. Some of the resources have already been exhausted while some of the resources have already futile due to extreme emergence of multiple drug resistance. But infections are nowhere close to containment. There are several reasons accountable for reemergence of any infectious disease-human behavior, noncompliance globalization travel and tourism environmental adaptation, and negligence of health.

By eradication of smallpox, scientists all over the world have gained huge win which formed the basis of future enthusiastic researches. There are many diseases in their final stages of eradication like polio and Dracunculiasis and these studies help still confidence that a complete elimination of any infection is definitely possible. This milestone however is still pending in the case of fungal infections. Hence, to achieve success, several eradication programs with international collaboration and public corporation are being run on a global level to contain fungal diseases.

Most people want to assume that they are relatively immune to the devastating effects of infectious diseases that affected previous generations. In the current circumstance, such reassurance is needed. The development of retroviruses as a global pandemic has already worsened the situation. But as current times have made us realize, new infectious diseases are always on the horizon. With the development of frequent travel and significant transport systems, new and severe infectious diseases are always a possibility. Initially these infections were categorized into outbreak infections and gradual infections where investigations were carried out to contain endemic and epidemic infections. These were mostly bacterial in nature, whereas, fungal infections have always been gradual parasitic and opportunistic in nature.

To, contain any infection, estimation of key parameters and analysis of transmission system are the very first step. These include gathering the information about occurrence in a population, transmission mode, natural history of infection, interactions between pathogen and host. With time, epidemiological data was given deserving importance and understanding of the epidemiology of any infections disease now became the most basic step carried in therapeutic research. Moreover, there are different ways to track a new infection: tracing mobile genetic elements, understanding phylogenetic relationships, identifying transmission systems, identifying specific pathogens which are accustomed to specific prospective host and identifying characteristics that lead to pathogenesis. Once the infection is identified and well characterized then clinical treatment and intervention strategies are executed. Along with treatment, rapid detection techniques for early diagnosis are of utmost importance.

When we talk about fungal infections, it is extremely important to distinguish between normal commensal and conventional pathogens. For instance, this book is about anticandidal drugs; so here we give a lot of emphasis on understanding candida and its infection. Candida is an opportunistic pathogen that only attacks immunosuppressed patients whereas there are thousands of fungi that directly infect on contact. Identification of pathogenic strain and pathogenic species is also very crucial. Not all the species of a genius will have pathogenic nature. Hence the identification of a causal agent an opportunistic agent in laboratory is rather difficult.

References

[1] R.I. Aminov, A brief history of the antibiotic era: lessons learned and challenges for the future, Frontiers in Microbiology 1 (2010) 134, https://doi.org/10.3389/fmicb.2010.00134.

[2] Y. Zhang, K. Zhao, Global eradication of small-pox: historical fact, experiences and enlightenment, Zhonghua Liu Xing Bing Xue Za Zhi 20 (1999) 67–70.

[3] K.I. Mohr, History of antibiotics research, Current Topics in Microbiology and Immunology 398 (2016) 237–272, https://doi.org/10.1007/82_2016_499.

[4] P. Manohar, B. Loh, R. Nachimuthu, X. Hua, S.C. Welburn, S. Leptihn, Secondary bacterial infections in patients with viral pneumonia, Frontiers in Medicine 7 (2020). https://www.frontiersin.org/article/10.3389/fmed.2020.00420. (Accessed 4 May 2022).

[5] E.C. Bonard, The plague and Alexander Yersin (1863–1943), Revue Medicale de la Suisse Romande 114 (1994) 389–391.

[6] M.A. Shampo, R.A. Kyle, Shibasaburo Kitasato-Japanese bacteriologist, Mayo Clinic Proceedings 74 (1999) 146, https://doi.org/10.4065/74.2.146.

[7] L. Pasteur, Chamberland, Roux, Summary report of the experiments conducted at Pouilly-le-Fort, near Melun, on the anthrax vaccination, 1881, The Yale Journal of Biology and Medicine 75 (2002) 59–62.

[8] K.A. Clark, R.M. Robinson, R.G. Marburger, L.P. Jones, J.H. Orchard, Malignant catarrhal fever in Texas cervids, Journal of Wildlife Diseases 6 (1970) 376–383, https://doi.org/10.7589/0090-3558-6.4.376.

[9] T.P. Monath, P.F.C. Vasconcelos, Yellow fever, Journal of Clinical Virology 64 (2015) 160–173, https://doi.org/10.1016/j.jcv.2014.08.030.

[10] D. Wright, Lyme disease, Journal of the American Academy of Nurse Practitioners 13 (2001) 223–226, https://doi.org/10.1111/j.1745-7599.2001.tb00024.x (quiz 227–228).

[11] A. Bluszcz-Roznowska, I. Olszok, E.J. Kucharz, Ehrlichiosis, Przeglad Lekarski 62 (2005) 1529–1531.

CHAPTER 3

Fungal diseases and antifungal drugs

Introduction

The majority of infections go unnoticed, but the infecting agents do sometimes evoke a reaction from the body, resulting in clinically apparent signs and symptoms, a condition known as infectious disease. Fungi became the most dangerous pathogen as methods to combat bacterial infections in patients improved. Yeasts and molds are now among the top 10 pathogens removed from patients in intensive care units [1]. Mycoses are fungal diseases that affect humans and are classified into four categories grounded on their degree of dissemination into body tissues:

Superficial mycoses

Fungi that develop only on the surface of the skin or hair trigger superficial infections. Dermatophytes or yeasts are responsible for the majority of superficial fungal infections of the skin. Fungal infections are rarely fatal, but they are often persistent or chronic in otherwise healthy people.

People with actual fungal infections have benefited from the availability of successful over-the-counter antifungal drugs, but these medications are often used by people with other skin diseases such as dermatitis. One of the most difficult aspects of diagnosing fungal infections is that they mimic dermatitis and other inflammatory conditions. Fungal infections are overdiagnosed and underdiagnosed by both doctors and patients. The choice of antifungal agent is dictated by the location and nature of the infection, as well as the causative organism, which is normally easily recognizable. *Onychomycosis*, for example, is normally treated with systemically available antifungals; terbinafine and itraconazole

CONTENTS

Introduction ... 33
Superficial mycoses 33
Cutaneous mycoses 34
Subcutaneous mycoses 34
Systemic or invasive mycoses 35
Candidal infection 38
Biofilm formation 39
Adhesion and invasins 39

Anticandidal Therapeutics. https://doi.org/10.1016/B978-0-443-18744-5.00002-2
Copyright © 2023 Elsevier Inc. All rights reserved.

Contact sensing and thigmotropism 40
Hydrolytic enzymes 41
pH sensing and regulation 41
Metabolic adaptation 43
Environmental stress response 43
Heat shock protein 44
Metal acquisition 45
Antifungal agents 45
Azoles 45
Polyenes 46
Echinocandins 47
Allylamines 48
Other drugs 48
Mechanism of antifungal resistance 49
Resistance to azoles..................... 49
Alteration of drug target 50
Upregulation of multidrug transporters 50
Cellular stress responses 51
Biofilms 52

Table 3.1 Superficial mycoses.

Infection name	Causative agent	References
Pityriasis versicolor Seborrhoeic dermatitis Follicular pityriasis	*Malassezia* spp.	[3]
Tinea nigra	*Hortaea werneckii*	[4]
White piedra	*Trichosporon* spp.	[5]
Black piedra	*Piedraia hortae*	[6]

accumulate in keratinous tissues, making them ideal agents for treating onychomycosis [2].

Cutaneous mycoses

Cutaneous mycoses are fungi that cause infections in the skin's keratinized layers and appendages like nail and hair. It has no impact on living cells. Dermatophytes and yeasts are the most common causes of these infections. Nondermatophyte molds, in addition to the recognized pathogens, are now becoming important in cutaneous mycoses [7]. As the host reacts to metabolic byproducts, the species colonize the keratin tissues, causing inflammation. Since dermatophytes are unable to penetrate tissue of an immunocompetent host, their colonies are normally restricted to the nonliving layer of the epidermis [8]. A host response ranging from mild to extreme is elicited by invasion. *Trichophyton*, *Microsporum*, and *Epidermophyton* are the three genera that make up this closely related group of species. *Arthroderma*, *Ctenomyces*, *Lophophyton*, *Nannizzia*, *Guarromyces*, and *Paraphyton* are among the other genera [9] (Table 3.2).

Subcutaneous mycoses

They invade the subcutaneous, connective and bone tissue under the skin. On a global level, the most common form of subcutaneous mycosis is sporotrichosis, which affects gardeners and farmers who work with soil. The fungus *Sporothrix schenckii* causes a chronic infection that can take three different forms. The

Table 3.2 Cutaneous mycoses.

Cutaneous mycoses	Causative agent	References
Dermatophytosis	*Arthroderma, Lophophyton, Microsporum, Nannizzia, Trichophyton, Epidermophyton*	[10]
Dermatomycosis	Nondermatophyte molds *Neoscytalidium, Scopulariopsis*	[11]

cutaneous lymphatic type is described by the formation of a single pustule or nodule at the invasion site. The lymphatic distribution and the appearance of multiple subcutaneous lesions follow. Several, painless cutaneous or subcutaneous nodules may develop into ulcers or abscesses affecting the muscles, joints, bones, eyes, gastrointestinal system, mucous membranes, and nervous system in this disseminated form [12]. Other types of subcutaneous mycoses are caused by a variety of fungal species and are mainly found in the tropics and subtropics (Table 3.1). These conditions are known as chromomycosis (a chronic, slow-moving, destructive infection that affects many layers of the skin and results in abscessing granulomas). Treatment is complex, and surgical removal of the offending tissues is often required. Skin lesions can be the only symptom of a systemic fungal infection in some patients, and early identification of these lesions may help with early diagnosis and treatment. While *Candida*-based subcutaneous abscess followed by transient candidemia is prevalent in developing countries, it is a rare complication in immunocompromised patients with HIV/AIDS [13] (Table 3.3).

Systemic or invasive mycoses

Systemic mycoses are fungi that affect the internal organs. In the right conditions, fungi may enter the body through the lungs, the gut, the paranasal sinuses, or the skin [20]. The fungus will then spread through the bloodstream to various organs, including

Modification of ergosterol biosynthetic pathway................52
Chromosomal alterations 52
Resistance to polyenes.........52
Resistance to echinocandins 53
Upregulation of multidrug transporters ..53
Drug efflux.....54
References 55

Table 3.3 Subcutaneous mycoses.

Infection	Pathogen	References
Sporotrichosis	*Sporothrix schenckii*	[14]
Zygomycosis	*Basidiobolus ranarum* *Conidiobolus coronatus* *Rhizomucor,* *Mucor,* *Cunninghamella,* *Saksenaea,* *Apophysomyces,* *Cokeromyces Mortierella.*	[15]
Chromoblastomycosis	*Fonsecaea pedrosoi,* *Fonsecaea compacta,* *Cladosporium carrionii,* *Phialophora verrucosa*	[16]
Phaeophycomycosis	*Exophiala jeanselmei* *Phialophora verrucosa* *Aureobasidium pullulans* *Cladosporium,* *Curvularia* *Alternaria* *Bipolaris* *Exserohilum* *Cladophialophora* *Verruconis*	[17]
Rhinosporidiosis	*Rhinosporidium seeberi*	[18]
Lobomycosis	*Lacazia loboi*	[18]
Mycetoma	*Madurella,* *Acremonium,* *Pseudallescheria,* *Exophiala,* *Leptosphaeria,* *Curvularia,* *Fusarium,* *Aspergillus*	[19]

the skin, causing multiple organ failure and ultimately the patient's death. Changes in medical and surgical care, especially in intensive care units that use invasive catheters for monitoring, as well as the use of more active immunosuppressive and antibiotic agents has resulted in a significant increase in the occurrence and variation of invasive fungal infections over the last 30 years. The high death rate caused by systemic mycoses is due to the delay in diagnosis time that not only increases the risk but also the severity of infections [21].

At the turn of the century, fungi that cause systemic mycoses have become a major concern, as well as a clinical and pharmaceutical problem, affecting people with weakened immune systems. Despite the fact that mycoses infect billions of people all over the world and destroy more than two million people per year, their effect has been underestimated [22]. Owing to their complicated differential diagnosis, fungal infections are either not identified or underreported in general (Table 3.4).

Table 3.4 Systemic infections.

Infection	Causal organism	Nature	References
Penicilliosis	Penicillium marneffei	Endemic	[23]
Histoplasmosis	Histoplasms capsulatum	Endemic	[24]
Blastomycosis	Blastomyces dermatitidis	Endemic	[25]
Coccidioidomycosis	Coccidioides immitis	Endemic	[26]
Paracoccidioidomycosis	Paracoccidioides brasiliensis	Endemic	[27]
Mucormycosis	Mucor spp Rhizopus spp Absidia spp	Opportunistic	[28]
Aspergillosis	Aspergillus flavus Aspergillus fumigatus	Opportunistic	[29]
Candidiasis	Candida albicans Candida tropicalis Candida glabrata Candida parapsilosis Candida krusei	Opportunistic	[30]
Cyptococcosis	Cryptococcus neoformans	Opportunistic	[31]

Candidal infection

Candida species are commensal fungi that live in a variety of host niches, comprising the mouth, GIT, vaginal cavity, and epidermal layer of skin. *Candida* may, however, cause infections, or candidiasis, in some cases, ranging from minor mucous membrane infections to life-threatening systemic diseases. However, *Candida albicans* is the most common fungal pathogen in life-threatening nosocomial blood infections in humans. Their causal infections are a serious clinical issue, particularly in HIV/AIDS patients, chemotherapy patients, and organ transplant recipients [32]. Furthermore, in the coming decades, there will be a significant rise in the number of elderly people who are vulnerable to fungal infections. Antifungal agents aren't always effective against *Candida albicans*, which is still a common pathogen in hospital-acquired infections, causing serious mucosal infections like oral candidiasis, onychomycosis, vulvovaginal candidiasis, and systemic mycoses, all of which have high mortality rates [33].

For young researchers, it is important to understand the basic question—what causes a normal commensal fungal to morph into a fatal infection causing pathogen? The answer lies in the complex interplay of internal proteins and external factors. A broad variety of virulence factors and fitness attributes help *C. albicans*' ability to infect such diverse host niches. The morphological dimorphic shift, cell surface proteins, thigmotropism, the development of biofilms, phenotypic switching, and the secretion of hydrolytic enzymes are all considered virulence factors [34]. Rapid adaptation to changes in environmental pH, metabolic stability, efficient nutrient acquisition systems, and effective stress response machines are also fitness attributes. *C. albicans* is a morphologically diverse fungus that can develop as ovoid-shaped budding yeast, elongated ellipsoid cells with septa (pseudohyphae), or true hyphae with parallel walls. White and opaque cells produced during switching, as well as chlamydospores, thick-walled sporelike structures, are other morphologies of this fungal species [35]. While morphological forms like yeast and true hyphae are commonly seen and studied, the exact function of pseudohyphae and switching in vivo is scientifically unclear [36].

Biofilm formation

Biofilm formation is the most studied process of fungal infection and growth and they are also essential in clinical infections, particularly because they are associated with high antibiotic resistance [37]. While biofilm formation is a process that occurs in all of the *Candida* species discussed above, it varies considerably from one species to another, as well as in its reliance on surface, host niche, and other factors. *C. albicans* mature biofilms, for example, have a more heterogeneous composition, with blastospores and hyphae surrounded by a polysaccharide-based ECM [37]. The ECM serves as a structural scaffold for cell adhesion to various surfaces and as a buffer between biofilm cells and the surrounding environment. Water channels normally surround the microcolonies within the structure of these biofilms In the case of *C. glabrata*; the biofilm is entirely made up of yeast type cells packed closely together in a multilayer system or in clusters of cells. In contrast, the biofilm of *C. tropicalis* resembles a network of yeast, pseudohyphae, and hyphae with strong hyphal budding, while the biofilm of *C. parapsilosis* resembles clusters of yeast cells adhered to the surface with limited ECM [38]. These variations illustrate the complexities of the processes that lead to biofilm formation, as well as the challenge of finding a single solution to eliminate all *Candida* biofilms.

Adhesion and invasins

C. albicans possesses a wide range of contributing factors that may be responsible for its pathogenesis. After initial adherence, the yeast-to-hypha transformation is triggered by contact with the host cell, and the hypha is guided to expand by thigmotropism [39]. Invasins, adhesion, and physical forces aid invasion into the host cell, which occurs by endocytosis or active penetration. There's a connection between adherence and the ability to colonize and spread disease [40]. *Candidia albicans*, for example, is strongly adherent and has been associated to a wide range of diseases, while *Candida krusei* and *Candida glabrata* adhere less well and are less commonly linked to disease. Due to the mannosylated surface proteins that cover the fungal cells, one of *C. albicans*' most significant virulence characteristics is cell surface hydrophobicity. Some of these proteins help the yeast bind to the host cells by increasing hyphal types and providing resistance to macrophages, both of which are essential for the development of chronic lesions. Colonization requires fungal adhesion to the epithelium, and infections are marked by the invasion of vaginal epithelial cells [41].

Contact sensing and thigmotropism

Many obligate plant pathogenic fungi use thigmotropism (the property to detect and respond to changes in surface contours) to locate epidermal entry sites and initiate invasive development [42]. Contact sensing is an essential environmental signal that causes *C. albicans* to form hyphae and biofilms [43]. Yeast cells turn to hyphal growth when they come into contact with a surface. These hyphae may then enter the substratum on some substrates, such as agar or mucosal surfaces. Biofilms are also formed when liquids come into contact with solid surfaces, which make the phenomenon of contact sensing a versatile factor. Hyphae in *C. albicans* shift their growth path in response to changes in surface topography (thigmotropism) and electrical fields (galvanotropism) [44]. External calcium ions are needed for these responses, and mutants with defects in calcium uptake and calcium signaling have reduced thigmotropic and galvanotropic responses. Physical contact with semisolid substrates, in addition to these tropisms, causes *C. albicans* hyphae to invade the substrates. Thus, *C. albicans* hyphal cells react to asymmetric external cues by orienting polarised growth, but the molecular mechanism that aligns the growth axis is unknown [45].

Initially some researches had demanded the need to understand the hyphal invasion in detail, to understand whether it is thigmotropism or chemotropism [46]. But later studies have found that directional hyphal growth (thigmotropism) may occur on surfaces with specific topologies like presence of ridges [47]. Brand et al. demonstrated that extracellular calcium uptake takes place through the calcium channels and Cch1, Mid1, and Fig. 3.1 regulate thigmotropism in *C. albicans* hyphae [48]. In addition, Brand et al. found that *C. albicans* thigmotropism is needed for complete epithelial cell damage and normal virulence in mice [48].

Hydrolytic enzymes

One element that plays a key role in the virulence process is hydrolytic activity, which contributes in the pathogenicity of bacteria, protozoa, and pathogenic yeasts. This can be categorized into extracellular proteinases and secreted aspartyl proteinases. Although several microorganisms have a range of hydrolytic enzymes, proteinases are by far the most frequently associated with virulence. Extracellular proteolytic activity, which is provided by a family of 10 secreted aspartyl proteinases, is critical for *Candida* pathogenicity (Sap proteins) [49]. Hydrolytic enzymes are virulence factors that the fungus secretes extracellularly. Saps are the most well-known hydrolytic enzymes formed by *Candida albicans*. Numerous studies have looked into the function of these Saps in *C. albicans* infections, but the physiological role of secreted phospholipases (PL) is unknown, and the role of lipases in virulence is also unknown [50].

pH sensing and regulation

C. albicans' ability to adapt and proliferate in a wide variety of host environments is one of the characteristics that have made it such a competitive opportunistic pathogen. Ambient pH is one of the most significant environmental variables that vary across niches. *C. albicans* can develop in a pH range of 2−10, and it has been isolated from a variety of anatomical sites with a wide range of ambient pH, including the stomach (pH 2), vagina (pH 4−5), and oral mucosa (pH 6), implying that tolerance to environmental pH is essential for *C. albicans* pathogenicity [51]. Its

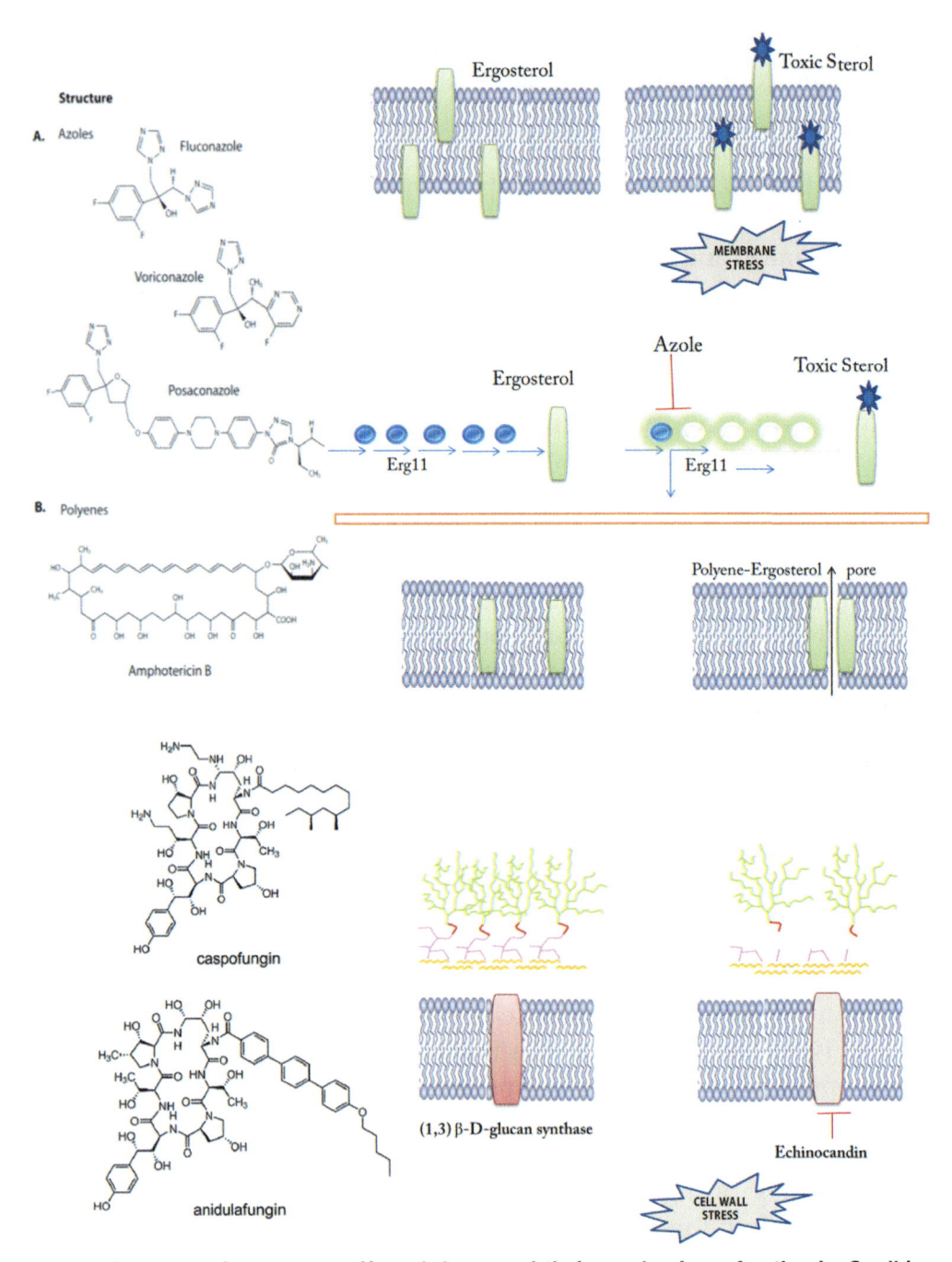

CHAPTER 3: Fungal diseases and antifungal drugs

FIGURE 3.1 Structure of common antifungal drugs and their mechanism of action in *C. albicans*: (A). Azoles affect ergosterol biosynthesis by targeting *ERG11* gene thereby accumulating toxic 14α-methylated sterols, (B). Polyenes form complex with sterols resulting in pores formation altering membrane permeability, (C). Echinocandins inhibits (1,3) β-D-glucan synthase damaging the cell wall.

adaptation to acidic conditions controls essential biological processes such as morbidity and mortality. Microorganisms adapt to a broad variety of pH by altering the expression of a large number of genes. Adaptive gene regulation induced by ambient pH is regulated by conserved signal transduction machinery in yeast and filamentous fungi. These pH signaling pathways have received a lot of attention [52].

Metabolic adaptation

Nutrients are needed by all living things, including fungi. *Candida* species must have a high level of metabolic flexibility and metabolic adaptation mechanisms in order to be effective in nutrient acquisition because they often occupy host niches with varying nutrient availability [53]. *C. albicans* uses aggressive penetration by hyphae extension and induced endocytosis as invasion mechanisms. Metabolism creates the precursors and energy needed for cell wall biosynthesis, antioxidant development, macromolecular repair, and protein refolding, to name a few [54]. Due to the importance of metabolism, antifungal drug therapies could target fungal-specific pathways or key enzymes with fungal-specific catalytic mechanisms [55].

Environmental stress response

Changes in atmospheric temperature, pH, osmolarity, and redox status are all thought to have affected the emergence of ancient life forms in various ecosystems. As a result, essential elements of cellular responses to these stresses can be found in all kingdoms of life [56]. Stress proteins are a group of proteins that prokaryotic and eukaryotic cells produce in response to different types of physiological stress. These proteins may be stress-induced or constitutive proteins whose development is significantly increased in response to stress [57]. In addition, the main regulators that control the heat shock response are conserved throughout the fungal kingdom and eukaryotes in general, and this evolutionary conservation of key regulatory modules is also true for other stress responses such as osmotic and oxidative stress [58].

Heat shock protein

Heat shock proteins (HSPs), also known as stress proteins, are a family of proteins found in all cells of all living things. When a cell is exposed to different forms of environmental pressures, such as heat, cold, or oxygen deprivation, they are caused. Heat shock proteins can be found in cells under normal circumstances [59]. They serve as "chaperones," ensuring that the cell's proteins are in the correct form and at the correct time. HSPs aid in the folding of new or twisted proteins, which is essential for their work. They also transport old proteins to "garbage disposals" within the cell and shuttle proteins from one compartment to another within the cell. Heat shock proteins are also believed to play a role in the presentation of pieces of proteins (or peptides) on the cell surface to help the immune system recognize diseased cells [60].

Hsps are abundant in *C. albicans* and play a role in a variety of cellular pathways, including calcium-calcineurin, MAPK, Ras1-cAMP-PKA, and cell cycle control signaling. Their associated proteins are used in many signaling molecules in these pathways [61]. Furthermore, several studies have shown that Hsps confer antifungal drug resistance in *C. albicans* by controlling these signaling pathways. As a result, pharmacological or genetic targeting of Hsps may improve *C. albicans* sensitivity to antifungal drugs and minimize pathogenicity [62]. These molecular chaperones with a wide range of functions within species are generally categorized based on their molecular size. Six types of Hsps have been described in *C. albicans*, each with a different molecular size. Hsp104, Hsp90, Hsp70, and Hsp60 are high molecular mass Hsps that require adenosine triphosphate (ATP) [63]. Hsp12 and Hsp21, the other two, are ATP-independent low molecular mass Hsps that range in size from 12 to 42 kDa [64]. Thus it is evident that the majority of these Hsps play an important role in *C. albicans* growth and virulence.

Metal acquisition

Iron acquisition factors and pathways are among the fitness and virulence attributes that enable fungal cells to obtain this important mineral in the host's iron-deficient environment. It gets micronutrients, such as metals, from its host, and metal bioavailability varies a lot at the host—pathogen interface [65]. In a mechanism known as "nutritional immunity," the host withholds metals such as Fe, Zn, and Mn from invading microbes as part of the innate immune response. *C. albicans* is well-equipped to deal with the constraints placed on these micronutrients, and it can activate a variety of pathways for scavenging Fe and Zn from host sources [66]. Several metal-regulated or metal-regulatory genes, including *PRA1*, *FTR1*, *CSR1*, and *HAP43*, have been identified as virulence factors, and their loss of function in animal models reduces virulence [67,68].

Antifungal agents

The discovery of the antifungal activities of griseofulvin by Oxford in 1939 was one of the first milestones in the production of active and safe antifungal agents. Patients with acute *Candida* infections and those who are immunocompromised can benefit from systemic antifungals. In patients with normal immune function, topical antifungals are used to treat clear, localized candidiasis. These antifungals could be manufactured synthetically or obtained from natural sources. Excessive use of synthetic antifungals, on the other hand, induces liver damage, allergic reactions, altered estrogen levels, and drug resistance. Antifungal medications are classified by their source, structure, and mode of action. An ideal antifungal should be one that has little or no toxicity to their human host; therefore, they should target pathways or biomolecules specific to fungi. Major classes of antifungal drugs that are vastly used against the fungal infection include (Fig. 3.1A—C).

Azoles

The first report of N-substituted imidazoles was published in 1960s presenting their role as antifungal agents [69]. They consist of five-membered nitrogen containing heterocyclic compounds including both imidazoles and trizoles. The compounds miconazole and econazole and then followed by ketoconazole, fluconazole, and itraconazole were proven to be useful in treating *Candida* infection. The clinical efficacy and safety resulted in the popularity of fluconazole. They are most widely used drugs against fungal infections from almost past two decades [70].

The predominant mechanism of action of these compounds has been shown to be the disruption of sterol biosynthesis through the inhibitory action of P450 14-demethylase (P45014dm) (encoded by *ERG11*) that occurs after the stoichiometric interaction of the N-3 (imidazole) or N-4 (triazoles) substituents of the azole ring with the haem ring of the P45014dm. In the absence of azole, this monooxygenase catalysis, the oxidative removal of the C-14 methyl group from the sterol molecule (lanosterol or eburicol) takes place[71]. This demethylation is accomplished through three sequential hydroxylation reactions which have been shown to be essential for the generation of a sterol molecule that can correctly orient within the phospholipid bilayer and thus fulfill the sterols' regulatory function in membrane fluidity and stability [72]. While the triazoles such as fluconazole appear to target 14-DM almost exclusively, the imidazoles such as ketoconazole have a broader spectrum of action and also act on various monooxygenases. In most fungal cells, ergosterol is the sterol best suited to maintain membrane integrity and activity [73]. A complete block of ergosterol synthesis is known to block the cell growth and affects the physical state of the membrane.

Polyenes

Polyenes are amphipathic drugs with both hydrophobic and hydrophilic groups that work by forming drug-lipid complexes with ergosterol that form a membrane-spanning channel by intercalating into the fungal cell membrane. This causes cellular ions like K^+ ions to leak out of the cell, causing the proton gradient to be broken. Many *Candida* species, as well as *Candida neoformans* and *Aspergillus* species, have in vitro and in vivo activity against the polyene amphotericin B [74]. While polyenes were used in clinical practice for more than 50 years, their main drawback was host toxicity, such as renal dysfunction, which is possibly caused by structural parallels between ergosterol and cholesterol in mammalian plasma membranes.

Polyenes macrolide antibiotics (amphotericin B and nystatin), produced by species of Streptomyces, are characterized by large lactone rings, containing three to eight conjugated double bonds, which are generally combined with the sugar moiety. The toxicity of amphotericin B to fungal cells is due to binding of this drug to sterols incorporated in the cell membranes changing as such the physical state of the membranes. This causes an impairment of membrane functions resulting in an enhanced permeability to protons and leakage of internal constituents such as K^+, Ca^{2+}, and PO_4^{3-}. Clinically useful polyenes—nystatin, amphotericin B (AmB), and primaricin—have a higher affinity for ergosterol than its mammalian counterpart, cholesterol, which limits their toxicity in mammalian cells [75]. Membrane permeability can also increase upon the addition of AmB to phospholipids in the absence of sterols. It has also been shown that oxidative processes are involved in AmB's mechanism of action; AmB can cause oxidative damage to fungal cell membrane and can produce many immunoadjuvant effects. However, there are evidences that AmB has an antioxidant effect in vivo, which protects fungal cells against oxidative attack from the host.

Echinocandins

Echinocandins are large lipopeptide molecules that noncompetitively inhibit synthesis of $(1-3)$-β-D-glucan; it results in loss of cell wall integrity and severe stress on cell wall without affecting the levels of mannan synthesis or nucleic acid [76]. They are reported to have antifungal property both in vivo and in vitro [77].

Inhibition of glucan synthase results in cytological changes characterized by pseudohyphae formation, thickening of cell wall and bud, and failure in separation from the mother cell. This is comparatively newer class of antifungal drug; there are three drugs in clinical use belonging to this class namely: caspofungins, micafungins, and anidulafungins.

Allylamines

Allylamines, such as terbinafine and naftifine, are a new category of membrane sterol biosynthesis inhibitors that are functionally and chemically different from the other main groups of antifungal drugs that target ergosterol biosynthesis [78]. Terbinafine is highly effective against dermatophytes in vivo and in vitro. Furthermore, preliminary evidence from Ghannoum's group and from Ryder and coworkers indicates that terbinafine has good activity against at least some azole-resistant *C. albicans* strains [79]. Allylamine inhibits early steps of ergosterol biosynthesis pathway mainly at squalene epoxidation step. The antifungal activity of the allylamines (naftifine and terbinafine), of the benzylamine derivative butenafine and the thiocarbamates (tolnaftate and tolciclate) results from their inhibition of squalene epoxidation by the enzyme squalene epoxidase (which is a product of *ERG1* gene causing deficiency of ergosterol and accumulation of squalene.

Other drugs

5-Flucytosine: 5-Flucytosine (5-FC) is transported into fungal cells by a cytosine permease, which also takes up adenine, guanine, hypoxanthine, and cytosine. 5-FC is fungus specific since mammalian cells have little or no cytosine deaminase. Once inside the cell, flucytosine is deaminated to 5-fluorouracil (5-FU) by cytosine deaminase. Since 5-FU is toxic to humans and has a low absorption by fungi, it cannot be used as an antifungal drug. 5-FU is broken down into 5-fluorouridine monophosphate (5-FUMP) and 5-fluorodeoxyuridine monophosphate (5-FDUMP) (5-FdUMP). The aforementioned is a precursor of aberrant RNA, while the latter is a thymidylate synthetase and thus DNA synthesis inhibitor.

1, 3β—Glucan synthase inhibitors: Pneumocandin and echinocandin derivatives act on β-(1, 3)-glucan synthase in yeasts and some molds, and have fungicidal activities on these species. In *S. cerevisiae*, β-(1, 3)-glucan synthase is a multienzyme complex with two subunits encoded by the *FKS1* and *FKS2* genes. Inhibition of this enzyme results in cytological and ultrastructural changes in fungi characterized by growth of pseudohyphae, thickened cell wall, and buds failing to separate from mother cells. Deletion of both genes in this yeast results in a lethal phenotype.

Morpholines: N-substituted morpholine fungicides (fenpropimorph and amorolfine) are active against fungi pathogenic to plants and humans. Morpholines inhibit Δ14-reductase (encoded by *ERG24*) and Δ8-Δ7 isomerase (encoded by *ERG2*). Inhibition of both enzymes results in a decreased ergosterol biosynthesis. However, the therapeutic efficacies of morpholines are limited to superficial fungal infections.

Mechanism of antifungal resistance

Candidia albicans exerts different resistance mechanisms against antifungal drugs used to treat fungal infection. Antifungal resistance has been less widespread than bacterial resistance to antibiotics in general. However, drug treatment for fungal infections has failed in the past. A drug-resistant, pathogenic fungus is one that grows and produces clinical symptoms of disease in the presence of a drug at its maximum concentration at the infection site [80].

Resistance to azoles

In recent years the increase in incidence of life threatening fungal infection is accompanied with overuse of azoles as preventive measures for high risk individuals. The fungistatic existence of azoles has resulted in the appearance of azole resistance in clinical isolates. The most common mechanism of azole resistance in clinical isolates includes alterations of drug target Erg11p and the overexpression of multidrug transporters. Multiple stress signaling pathways also have been shown to be involved in acquired drug resistance [81,82]. In addition elevated drug resistance in *Candida* biofilms is also observed due to multiple pathways.

Alteration of drug target

The impact of an azole on *C. albicans* can be reduced by changing its target, which can be done by point mutations or overexpression. One of the most common mechanisms for conferring resistance is a mutation in the azole target Erg11p, which lowers the drug binding affinity [83]. The loss of heterozygosity, which results in the substitution of the wild type allele with the mutant allele, increases azole resistance even further [84]. Overexpression of Erg11p has been found in drug-resistant clinical isolates, although it is usually followed by other changes. Overexpression of Erg11p is possible thanks to a mutation in the transcription factor UPC2, which controls its expression. Upc2p binds to the azole-responsive enhancer factor (ARE), a region in the ERG11 promoter that is divided into two distinct 7-bp sequences at positions 224 and 251. The azole induction of ERG11 requires and is facilitated by both the ARE and Upc2p [85].

Furthermore, the G648D and G643A point mutations in Upc2p cause hyperactivation of the transcription factor, resulting in increased fluconazole resistance and overexpression of ergosterol biosynthesis genes [86]. A different mechanism to target overexpression is genomic changes that result in an increased gene dosage of ERG11 [87].

Upregulation of multidrug transporters

Overexpression of two essential ABC (ATP binding cassette) family transporter genes, CDR1 and CDR2, has resulted in azole resistance in clinical isolates of *C. albicans*. Expression of both these genes is shown to be regulated by transcription factor *TAC1* that binds to DRE (drug responsive element) in their promoters. In azole-resistant clinical isolates, a TAC1 hyperactive allele with an Asp977-to-Asn977 gain-of-function mutation has been discovered, conferring constitutive upregulation of CDR1 and CDR2 [84]. It has been shown that homozygosity at *TAC1* locus is associated to homozygosity at MTL locus, which in turn emphasize azole resistance in clinical isolates [88].

Some azole-resistant clinical isolates are independent of overexpression of ABC transporters (*CDR1* and *CDR2*); instead they exhibit overexpression of another transporter belonging to MFS (Major facilitator superfamily) transporters; *MDR1*. The multidrug resistance regulator (MRR1) is a transcription factor that regulates MDR1 expression and is upregulated in drug-resistant clinical isolates in tandem with MDR1. In fact, overexpression of MDR1 is often associated with a gain-of-function mutation in MRR1 in all clinical isolates [89]. Thus there is similarity of resistance mechanism that works together to help *C. albicans* survive in the presence of the azoles [90].

Cellular stress responses

In addition to above mentioned mechanism, *C. albicans* has evolved stress response mechanisms that enable cell to survive the stress conditions it encounters at various environmental niches. A mutation in the −5, 6-desaturase encoded by ERG3 is one well-characterized mechanism that decreases drug toxicity and confers resistance to stress responses [91]. This prevents the toxic sterol 14-methyl-3, 6-diol from accumulating when Erg11p is inhibited; instead, an alternative sterol, 14-methyl fecosterol, is introduced into the membrane, enabling the fungal cell to expand and differentiate in the presence of the azoles. This resistance mechanism has been discovered in clinical isolates [92].

Another key regulator of cellular stress response is Ca^{2+}-calmodulin-activated protein phosphatase calcineurin. It plays a critical role in tolerating membrane stress exerted by fluconazole treatment. To treat *C. albicans* infections, calcineurin inhibition with the well-known inhibitor tacrolimus (FK506) or cyclosporine works synergistically with fluconazole [93,94].

Biofilms

Biofilms are intricate architecture of different cell types covered within an extracellular matrix initiated by adherence of free-moving planktonic cells to surfaces, e.g., catheters or dentures. Biofilm has distinct phenotype in comparison to their planktonic counterpart. Biofilms also exhibit increased drug resistance but the mechanism is unclear it could partly be because of overexpression of transporters such as *CDR1*, *CDR2*, and *MDR1* also the levels of ergosterol are depleted in intermediate or mature biofilms [95].

Modification of ergosterol biosynthetic pathway

Azole resistance in *C. albicans* was earlier supposed to occur mainly through a modification or an overexpression of the *ERG 11*, which encode 14α-lanosterol demethylase (CYP 51, or P45014 DM) involved in sterol biosynthesis. Azoles bind to and inhibit P45014DM, which is involved in ergosterol biosynthesis in fungi [96]. When P45014DM is inhibited, a high level of methylated sterols is produced, causing membrane disruption. The most common resistant strategies used by fungal cells are point mutations in the target protein p45014DM and overexpression of the target protein's gene (*ERG* 11) [97].

Chromosomal alterations

Fungi are flexible with their genome as these can undergo karyotype variability. Alteration in chromosomal copy number as mechanism of regulating gene expression in *C. albicans* has been observed, where nondisjunction of two specific chromosomes is proposed as cause of in vitro development of fluconazole resistance.

Resistance to polyenes

Polyenes are another popular class of antifungal drug but due to their poor solubility and host toxicity their use was limited and therefore resistance of clinical isolates toward them is also rare. However some of *Candida* species display intrinsic resistance to amphotericin B (AMB) polyenes, such as *C. lusitaniae* and *C. guilliermondii* [98]. It has been documented that patients undergoing long-term FLC therapy exhibit defective C-5,6-desaturase, leading to block in ergosterol synthesis and accumulation of alternate sterol in the membrane and hence show cross-resistant to AMB [99].

Resistance to echinocandins

Echinocandins are relatively newer class of antifungal but resistance toward them has been reported in clinical isolates. Echinocandins are lipoprotein molecules that noncompetitively inhibits (1,3)-β-D-glucan synthase [100]. Biosynthesis and upkeep of the cell wall is a highly dynamic system that responds to a wide range of cell stresses and is closely regulated during cell growth and morphogenesis. It involves experimenting with existing biosynthetic machinery through interactions with cell stress and cell integrity pathways, as well as Golgi complex secretory vesicle delivery of new enzymes [101].

Upregulation of multidrug transporters

In contrast with azole upregulation of multidrug transporters have only limited role imparting resistance to echinocandins. Echinocandins such as micafungins and caspofungins are not substrate of *CDR1*, *CDR2*, and *MDR1*; therefore, regardless of strains overexpressing these pumps resistance toward azoles they do not show much resistance to echinocandins [102].

Calcineurin is key regulator of cellular stress; however, its role in resistance to echinocandins is unclear. Inhibition of calcineurin by cyclosporine does not show susceptibility to echinocandins except when concentration is highly increased [103]. Their involvement at basal expression level is not proven in echinocandins resistance. Inhibition of calcineurin acts synergistically with echinocandins in some strains of *C. albicans*. It has been well established that compromising calcineurin function phenocopies compromising Hsp90p function. Calcineurin is client protein of

HSP90p in *C. albicans* and *HSP90* inhibition blocks calcineurin activation. Given the genetic compromise of *C. albicans*, Hsp90p's function in echinocandin resistance has therapeutic potential. In a murine model of disseminated candidiasis, HSP90 expression improved the efficacy of an echinocandin [104].

Modification and degradation of drugs: Alterations in drug processing are important drug resistance mechanisms in a variety of cell types. Until now, no evidence for degradation of azole antifungal compounds has been seen in yeasts.

Drug import: Hydrophobic nature of drugs is presumed to allow their easy import via passive diffusion through the lipid bilayer into the cell. Thus fluctuations in membrane fluidity are expected to affect passive diffusion and therefore sensitivity to drugs. It was observed in a study when ERG mutants of *S. cerevisiae* having defective ergosterol biosynthetic pathway were used. These show high membrane fluidity and hypersensitive to several drugs. The enhanced fluidity was linked to enhance to diffusion of drugs. Along with fluidity, mutations leading to change in lipid composition also influence passive diffusion of drug.

Drug efflux

Several azoles-resistant clinical isolates of *C. albicans* as well as other fungal pathogen like *Aspergillus fumigatus* and *Cryptococcus neoformans* display transcriptional activation of efflux pump encoding genes and often show reduce intracellular accumulation of drugs. These resistant isolates primarily overexpress genes encoding multidrug efflux transporter proteins of two super families; the ABC transporter and MFS transporter [105]. Among several mechanisms of resistance active efflux of drug is an important mechanism of drug resistance to azole antifungals. Efflux pumps, generally present in plasma membrane, have now emerged as one of the important component of multi drug resistances [106]. This chapter has discussed about fungal disease and antifungal drugs. The next chapter is explaining multidrug transporters of fungus *C. albicans*.

References

[1] B.E. de Pauw, What are fungal infections? Mediterranean Journal of Hematology and Infectious Diseases 3 (2011) https://doi.org/10.4084/MJHID.2011.001.

[2] J.F.G.M. Meis, P.E. Verweij, Current management of fungal infections, Drugs 61 (2001) 13–25, https://doi.org/10.2165/00003495-200161001-00002.

[3] S.K. Ghosh, S.K. Dey, I. Saha, J.N. Barbhuiya, A. Ghosh, A.K. Roy, Pityriasis versicolor: a clinicomycological and epidemiological study from a tertiary care hospital, Indian Journal of Dermatology 53 (2008) 182–185, https://doi.org/10.4103/0019-5154.44791.

[4] A. Bonifaz, H. Badali, G.S. de Hoog, M. Cruz, J. Araiza, M.A. Cruz, L. Fierro, R.M. Ponce, Tinea nigra by Hortaea werneckii, a report of 22 cases from Mexico, Studies in Mycology 61 (2008) 77–82, https://doi.org/10.3114/sim.2008.61.07.

[5] M. Karray, W.P. McKinney, Tinea versicolor, in: StatPearls, StatPearls Publishing, Treasure Island (FL, 2021. http://www.ncbi.nlm.nih.gov/books/NBK482500/. (Accessed 14 November 2021).

[6] P. Sharma, A. Nassereddin, S. Sonthalia, Black Piedra, in: StatPearls, StatPearls Publishing, Treasure Island (FL, 2021. http://www.ncbi.nlm.nih.gov/books/NBK545178/. (Accessed 14 November 2021).

[7] D.W. Warnock, 61 - Fungi: superficial, subcutaneous and systemic mycoses, in: D. Greenwood, M. Barer, R. Slack, W. Irving (Eds.), Medical Microbiology, eighteenth ed., Churchill Livingstone, Edinburgh, 2012, pp. 616–641, https://doi.org/10.1016/B978-0-7020-4089-4.00075-5.

[8] T.J. Walsh, D.M. Dixon, Spectrum of mycoses, in: S. Baron (Ed.), Medical Microbiology, fourth ed., University of Texas Medical Branch at Galveston, Galveston (TX), 1996. http://www.ncbi.nlm.nih.gov/books/NBK7902/. (Accessed 20 May 2021).

[9] C.C. Ekwealor, C.A. Oyeka, Cutaneous mycoses among rice farmers in Anambra state, Nigeria, Journal of Mycology 2013 (2013) e190742, https://doi.org/10.1155/2013/190742.

[10] I. Weitzman, R.C. Summerbell, The dermatophytes, Clinical Microbiology Reviews 8 (1995) 240–259.

[11] D.T. Lakshmipathy, K. Kannabiran, Review on dermatomycosis: pathogenesis and treatment, Natural Science 2 (2010) 726–731, https://doi.org/10.4236/ns.2010.27090.

[12] M. Corti, M.F. Villafane, Fern, O. Messina, R. Negroni, Subcutaneous abscess as a single manifestation of candidiasis, Medical Mycology: Open Access 1 (2015), https://doi.org/10.21767/2471-8521.100006.

[13] F.F. Tuon, A.C. Nicodemo, Candida albicans skin abscess, Revista Do Instituto de Medicina Tropical de São Paulo 48 (2006) 301–302, https://doi.org/10.1590/S0036-46652006000500012.

[14] R. Orofino-Costa, P.M. de Macedo, A.M. Rodrigues, A.R. Bernardes-Engemann, Sporotrichosis: an update on epidemiology, etiopathogenesis, laboratory and clinical therapeutics, Anais Brasileiros de Dermatologia 92 (2017) 606–620, https://doi.org/10.1590/abd1806-4841.2017279.

[15] M.M. Roden, T.E. Zaoutis, W.L. Buchanan, T.A. Knudsen, T.A. Sarkisova, R.L. Schaufele, M. Sein, T. Sein, C.C. Chiou, J.H. Chu, D.P. Kontoyiannis, T.J. Walsh, Epidemiology and outcome of zygomycosis: a review of 929 reported cases, Clinical Infectious Diseases 41 (2005) 634–653, https://doi.org/10.1086/432579.

[16] R. Agarwal, G. Singh, A. Ghosh, K.K. Verma, M. Pandey, I. Xess, Chromoblastomycosis in India: review of 169 cases, PLoS Neglected Tropical Diseases 11 (2017) e0005534, https://doi.org/10.1371/journal.pntd.0005534.

[17] V. Sundaramoorthy, F. Duarte, P.M. Raj, J.S. Michael, P. Rupali, Phaeohyphomycosis: a 10-year review (2006–2016), Open Forum Infectious Diseases 4 (2017) S86, https://doi.org/10.1093/ofid/ofx163.038.

[18] S. Das, B. Kashyap, M. Barua, N. Gupta, R. Saha, L. Vaid, A. Banka, Nasal rhinosporidiosis in humans: new interpretations and a review of the literature of this enigmatic disease, Medical Mycology 49 (2011) 311–315, https://doi.org/10.3109/13693786.2010.526640.

[19] R.F. Omer, N. Seif EL Din, F.A. Abdel Rahim, A.H. Fahal, Hand mycetoma: the mycetoma Research Centre experience and literature review, PLoS Neglected Tropical Diseases 10 (2016) e0004886, https://doi.org/10.1371/journal.pntd.0004886.

[20] D.H. Van Thiel, M. George, C.M. Moore, Fungal infections: their diagnosis and treatment in transplant recipients, International Journal of Hepatology 2012 (2012) e106923, https://doi.org/10.1155/2012/106923.

[21] K.H. Neppelenbroek, R.S. Seó, V.M. Urban, S. Silva, L.N. Dovigo, J.H. Jorge, N.H. Campanha, Identification of Candida species in the clinical laboratory: a review of conventional, commercial, and molecular techniques, Oral Diseases 20 (2014) 329–344, https://doi.org/10.1111/odi.12123.

[22] A. Kühbacher, A. Burger-Kentischer, S. Rupp, Interaction of Candida species with the skin, Microorganisms 5 (2017), https://doi.org/10.3390/microorganisms5020032.

[23] J. Zheng, X. Gui, Q. Cao, R. Yang, Y. Yan, L. Deng, J. Lio, A clinical study of acquired immunodeficiency syndrome associated Penicillium

Marneffei infection from a non-endemic area in China, PLoS One 10 (2015) e0130376, https://doi.org/10.1371/journal.pone.0130376.

[24] M.A. Rahim, S. Zaman, M.R. Amin, K.N. Uddin, J.C. MA, Histoplasmosis: an emerging or neglected disease in Bangladesh? A systematic review, Oman Medical Journal 35 (2020) e91, https://doi.org/10.5001/omj.2020.09.

[25] J.A. McBride, G.M. Gauthier, B.S. Klein, Clinical manifestations and treatment of blastomycosis, Clinics in Chest Medicine 38 (2017) 435–449, https://doi.org/10.1016/j.ccm.2017.04.006.

[26] J. Brown, K. Benedict, B.J. Park, G.R. Thompson, Coccidioidomycosis: epidemiology, Clinical Epidemiology 5 (2013) 185–197, https://doi.org/10.2147/CLEP.S34434.

[27] R. Martinez, New trends in paracoccidioidomycosis epidemiology, Journal of Fungi 3 (2017) 1, https://doi.org/10.3390/jof3010001.

[28] D.P. Kontoyiannis, R.E. Lewis, O. Lotholary, B. Spellberg, G. Petrikkos, E. Roillides, A. Ibrahim, T.J. Walsh, Future directions in mucormycosis research, Clinical Infectious Diseases 54 (2012) S79, https://doi.org/10.1093/cid/cir886. –S85.

[29] J.-P. Latgé, G. Chamilos, Aspergillus fumigatus and aspergillosis in 2019, Clinical Microbiology Reviews 33 (2019) e00140–18, https://doi.org/10.1128/CMR.00140-18.

[30] F. Lamoth, S.R. Lockhart, E.L. Berkow, T. Calandra, Changes in the epidemiological landscape of invasive candidiasis, Journal of Antimicrobial Chemotherapy 73 (2018) i4–i13, https://doi.org/10.1093/jac/dkx444.

[31] E.K. Maziarz, J.R. Perfect, Cryptococcosis, Infectious Disease Clinics of North America 30 (2016) 179–206, https://doi.org/10.1016/j.idc.2015.10.006.

[32] R. Rautemaa, G. Ramage, Oral candidosis – clinical challenges of a biofilm disease, Critical Reviews in Microbiology 37 (2011) 328–336, https://doi.org/10.3109/1040841X.2011.585606.

[33] A. Jha, A. Kumar, Multiple drug targeting potential of novel ligands against virulent proteins of Candida albicans, International Journal of Peptide Research and Therapeutics 26 (2020), https://doi.org/10.1007/s10989-019-09897-1.

[34] F.L. Mayer, D. Wilson, B. Hube, Candida albicans pathogenicity mechanisms, Virulence 4 (2013) 119–128, https://doi.org/10.4161/viru.22913.

[35] M. Whiteway, C. Bachewich, Morphogenesis in Candida albicans, Annual Review of Microbiology 61 (2007) 529–553, https://doi.org/10.1146/annurev.micro.61.080706.093341.

[36] P.E. Sudbery, The germ tubes of Candida albicans hyphae and pseudo-hyphae show different patterns of septin ring localization, Molecular Microbiology 41 (2001) 19−31.

[37] J. Chandra, D.M. Kuhn, P.K. Mukherjee, L.L. Hoyer, T. McCormick, M.A. Ghannoum, Biofilm formation by the fungal pathogen Candida albicans: development, architecture, and drug resistance, Journal of Bacteriology 183 (2001) 5385−5394, https://doi.org/10.1128/JB.183.18.5385-5394.2001.

[38] M. Cavalheiro, M.C. Teixeira, Candida biofilms: threats, challenges, and promising strategies, Frontiers in Medicine 5 (2018) 28, https://doi.org/10.3389/fmed.2018.00028.

[39] D.L. Moyes, J.P. Richardson, J.R. Naglik, Candida albicans-epithelial interactions and pathogenicity mechanisms: scratching the surface, Virulence 6 (2015) 338−346, https://doi.org/10.1080/21505594.2015.1012981.

[40] H. Martin, K. Kavanagh, T. Velasco-Torrijos, Targeting adhesion in fungal pathogen Candida albicans, Future Medicinal Chemistry 13 (2020) 313−334, https://doi.org/10.4155/fmc-2020-0052.

[41] J.C.O. Sardi, C. Duque, F.S. Mariano, M.R. Marques, J.F. Höfling, R.B. Gonçalves, Adhesion and invasion of Candida albicans from periodontal pockets of patients with chronic periodontitis and diabetes to gingival human fibroblasts, Medical Mycology 50 (2012) 43−49, https://doi.org/10.3109/13693786.2011.586133.

[42] H. Nikawa, H. Nishimura, T. Hamada, S. Makihira, L.P. Samaranayake, Relationship between thigmotropism and Candida biofilm formation in vitro, Mycopathologia 144 (1998) 125−129, https://doi.org/10.1023/a:1007073930933.

[43] N.A. Gow, Germ tube growth of Candida albicans, Current Topics in Medical Mycology 8 (1997) 43−55.

[44] N.A.R. Gow, B. Yadav, Microbe profile: Candida albicans: a shape-changing, opportunistic pathogenic fungus of humans, Microbiology 163 (2017) 1145−1147, https://doi.org/10.1099/mic.0.000499.

[45] A. Brand, Hyphal Growth in Human Fungal Pathogens and, vol 2012, 2012, https://doi.org/10.1155/2012/517529.

[46] J.M. Davies, A.J. Stacey, C.A. Gilligan, Candida albicans hyphal invasion: thigmotropism or chemotropism? FEMS Microbiology Letters 171 (1999) 245−249, https://doi.org/10.1111/j.1574-6968.1999.tb13439.x.

[47] H. Nikawa, H. Nishimura, T. Hamada, S. Sadamori, Quantification of thigmotropism (contact sensing) of Candida albicans and Candida tropicalis, Mycopathologia 138 (1997) 13−19, https://doi.org/10.1023/a:1006849532064.

[48] A. Brand, N.A. Gow, Mechanisms of hypha orientation of fungi, Current Opinion in Microbiology 12 (2009) 350–357, https://doi.org/10.1016/j.mib.2009.05.007.

[49] M. Schaller, C. Borelli, H.C. Korting, B. Hube, Hydrolytic enzymes as virulence factors of Candida albicans, Mycoses 48 (2005) 365–377, https://doi.org/10.1111/j.1439-0507.2005.01165.x.

[50] J.R. Naglik, S.J. Challacombe, B. Hube, Candida albicans secreted aspartyl proteinases in virulence and pathogenesis, Microbiology and Molecular Biology Reviews 67 (2003) 400–428, https://doi.org/10.1128/MMBR.67.3.400-428.2003.

[51] T. Maeda, The signaling mechanism of ambient pH sensing and adaptation in yeast and fungi, The FEBS Journal 279 (2012) 1407–1413, https://doi.org/10.1111/j.1742-4658.2012.08548.x.

[52] S.L. Sherrington, E. Sorsby, N. Mahtey, P. Kumwenda, M.D. Lenardon, I. Brown, E.R. Ballou, D.M. MacCallum, R.A. Hall, Adaptation of Candida albicans to environmental pH induces cell wall remodelling and enhances innate immune recognition, PLoS Pathogens 13 (2017) e1006403, https://doi.org/10.1371/journal.ppat.1006403.

[53] A.J.P. Brown, G.D. Brown, M.G. Netea, N.A.R. Gow, Metabolism impacts upon Candida immunogenicity and pathogenicity at multiple levels, Trends in Microbiology 22 (2014) 614–622, https://doi.org/10.1016/j.tim.2014.07.001.

[54] S.Y. Chew, W.J.Y. Chee, L.T.L. Than, The glyoxylate cycle and alternative carbon metabolism as metabolic adaptation strategies of Candida glabrata: perspectives from Candida albicans and Saccharomyces cerevisiae, Journal of Biomedical Science 26 (2019) 52, https://doi.org/10.1186/s12929-019-0546-5.

[55] S.Y. Ting, O.A. Ishola, M.A. Ahmed, Y.M. Tabana, S. Dahham, M.T. Agha, S.F. Musa, R. Muhammed, L.T.L. Than, D. Sandai, Metabolic adaptation via regulated enzyme degradation in the pathogenic yeast Candida albicans, Journal of Medical Mycology 27 (2017) 98–108, https://doi.org/10.1016/j.mycmed.2016.12.002.

[56] A.J.P. Brown, D.E. Larcombe, A. Pradhan, Thoughts on the evolution of core environmental responses in yeasts, Fungal Biology 124 (2020) 475–481, https://doi.org/10.1016/j.funbio.2020.01.003.

[57] A.P. Gasch, Comparative genomics of the environmental stress response in ascomycete fungi, Yeast 24 (2007) 961–976, https://doi.org/10.1002/yea.1512.

[58] A.J.P. Brown, S. Budge, D. Kaloriti, A. Tillmann, M.D. Jacobsen, Z. Yin, I.V. Ene, I. Bohovych, D. Sandai, S. Kastora, J. Potrykus, E.R. Ballou, D.S. Childers, S. Shahana, M.D. Leach, Stress adaptation in a pathogenic

fungus, Journal of Experimental Biology 217 (2014) 144–155, https://doi.org/10.1242/jeb.088930.

[59] R.C.Y. Matthews, Candida albicans HSP 90: link between protective and auto immunity, Journal of Medical Microbiology 36 (1992) 367–370, https://doi.org/10.1099/00222615-36-6-367.

[60] M.L. Zeuthen, D.H.Y. Howard, Thermotolerance and the heat-shock response in Candida albicans, Microbiology 135 (1989) 2509–2518, https://doi.org/10.1099/00221287-135-9-2509.

[61] H. Huang, D. Harcus, M. Whiteway, Transcript profiling of a MAP kinase pathway in C. albicans, Microbiological Research 163 (2008) 380–393, https://doi.org/10.1016/j.micres.2008.03.001.

[62] J.P. Burnie, T.L. Carter, S.J. Hodgetts, R.C. Matthews, Fungal heat-shock proteins in human disease, FEMS Microbiology Reviews 30 (2006) 53–88, https://doi.org/10.1111/j.1574-6976.2005.00001.x.

[63] I.N. Cruz, Y. Zhang, M. de la Fuente, A. Schatzlein, M. Yang, Functional characterization of heat shock protein 90 targeted compounds, Analytical Biochemistry 438 (2013) 107–109, https://doi.org/10.1016/j.ab.2013.03.026.

[64] F.L. Mayer, D. Wilson, I.D. Jacobsen, P. Miramón, S. Slesiona, I.M. Bohovych, A.J.P. Brown, B. Hube, Small but crucial: the novel small heat shock protein Hsp21 mediates stress adaptation and virulence in Candida albicans, PLoS One 7 (2012) e38584, https://doi.org/10.1371/journal.pone.0038584.

[65] C.X. Li, J.E. Gleason, S.X. Zhang, V.M. Bruno, B.P. Cormack, V.C. Culotta, Candida albicans adapts to host copper during infection by swapping metal cofactors for superoxide dismutase, PNAS 112 (2015) E5336–E5342, https://doi.org/10.1073/pnas.1513447112.

[66] C. Simm, R.C. May, Zinc and iron homeostasis: target-based drug screening as new route for antifungal drug development, Frontiers in Cellular and Infection Microbiology 9 (2019), https://doi.org/10.3389/fcimb.2019.00181.

[67] D.H. Howard, Acquisition, transport, and storage of iron by pathogenic fungi, Clinical Microbiology Reviews 12 (1999) 394–404, https://doi.org/10.1128/CMR.12.3.394.

[68] S. Hameed, T. Prasad, D. Banerjee, A. Chandra, C.K. Mukhopadhyay, S.K. Goswami, A.A. Lattif, J. Chandra, P.K. Mukherjee, M.A. Ghannoum, R. Prasad, Iron deprivation induces EFG1-mediated hyphal development in Candida albicans without affecting biofilm formation, FEMS Yeast Research 8 (2008) 744–755, https://doi.org/10.1111/j.1567-1364.2008.00394.x.

[69] D.J. Sheehan, C.A. Hitchcock, C.M. Sibley, Current and emerging azole antifungal agents, Clinical Microbiology Reviews 12 (1999) 40–79.

[70] A.T. Nishimoto, C. Sharma, P.D. Rogers, Molecular and genetic basis of azole antifungal resistance in the opportunistic pathogenic fungus Candida albicans, Journal of Antimicrobial Chemotherapy 75 (2020) 257–270, https://doi.org/10.1093/jac/dkz400.

[71] A. Taraszkiewicz, G. Szewczyk, T. Sarna, K.P. Bielawski, J. Nakonieczna, Photodynamic inactivation of Candida albicans with imidazoacridinones: influence of irradiance, photosensitizer uptake and reactive oxygen species generation, PloS One 10 (2015) e0129301, https://doi.org/10.1371/journal.pone.0129301.

[72] V.M. Chiocchio, Determination of ergosterol in cellular fungi by HPLC. A modified technique, Journal of the Argentine Chemical Society 98 (2011) 10–15.

[73] R. Prasad, M.A. Ghanoum, Revival: Lipids of Pathogenic Fungi, CRC Press, 1996. 302 p. ISBN 9781138560581. January 25, 2019

[74] L. Ostrosky-Zeichner, K.A. Marr, J.H. Rex, S.H. Cohen, S.H. Cohen, Amphotericin B: time for a new "gold standard", Clinical Infectious Diseases: An Official Publication of the Infectious Diseases Society of America 37 (2003) 415–425, https://doi.org/10.1086/376634.

[75] N.H. Georgopapadakou, T.J. Walsh, Antifungal agents: Chemotherapeutic targets and immunologic strategies, Antimicrobial Agents and Chemotherapy 40 (2) (1996) 279–291.

[76] D.W. Denning, Echinocandins: a new class of antifungal, Journal of Antimicrobial Chemotherapy 49 (2002) 889–891, https://doi.org/10.1093/jac/dkf045.

[77] D. Andes, D.J. Diekema, M.A. Pfaller, J. Bohrmuller, K. Marchillo, A. Lepak, In vivo comparison of the pharmacodynamic targets for echinocandin drugs against Candida species, Antimicrobial Agents and Chemotherapy 54 (2010) 2497–2506, https://doi.org/10.1128/AAC.01584-09.

[78] A. Kumar, A. Jha, Anticandidal Agents, Academic Press, 2016.

[79] A. Butts, D.J. Krysan, Antifungal drug discovery: something old and something new, PLoS Pathogens 8 (2012) 9–11, https://doi.org/10.1371/journal.ppat.1002870.

[80] G. Molero, F. Navarro-garcía, M. Sánchez-pérez, Candida albicans: genetics, dimorphism and pathogenicity, International Microbiology 1 (1998) 95–106.

[81] B.R. Braun, A.D. Johnson, TUP1, CPH1 and EFG1 make independent contributions to filamentation in candida albicans, Genetics 155 (2000) 57–67.

[82] B. a Arthington-skaggs, H. Jradi, T. Desai, C.J. Morrison, Quantitation of ergosterol content: novel method for determination of fluconazole

susceptibility of Candida albicans, Susceptibility of Candida albicans 37 (1999) 3332–3337.

[83] K.E. Pristov, M.A. Ghannoum, Resistance of Candida to azoles and echinocandins worldwide, Clinical Microbiology and Infection 25 (2019) 792–798, https://doi.org/10.1016/j.cmi.2019.03.028.

[84] A.T. Coste, M. Karababa, F. Ischer, J. Bille, D. Sanglard, TAC1, transcriptional activator of CDR genes, is a new transcription factor involved in the regulation of Candida albicans ABC transporters CDR1 and CDR2, Eukaryotic Cell 3 (2004) 1639–1652, https://doi.org/10.1128/EC.3.6.1639-1652.2004.

[85] A.A. Sagatova, M.V. Keniya, R.K. Wilson, M. Sabherwal, J.D.A. Tyndall, B.C. Monk, Triazole resistance mediated by mutations of a conserved active site tyrosine in fungal lanosterol 14α-demethylase, Scientific Reports 6 (2016) 26213, https://doi.org/10.1038/srep26213.

[86] S.J. Hoot, X. Zheng, C.J. Potenski, T.C. White, H.L. Klein, The role of Candida albicans homologous recombination factors Rad54 and Rdh54 in DNA damage sensitivity, BMC Microbiology 11 (2011) 214, https://doi.org/10.1186/1471-2180-11-214.

[87] L.E. Cowen, J.B. Anderson, L.M. Kohn, Evolution of drug resistance in Candida albicans, Annual Review of Microbiology 56 (2002) 139–165, https://doi.org/10.1146/annurev.micro.56.012302.160907.

[88] N.A. Gaur, R. Manoharlal, P. Saini, T. Prasad, G. Mukhopadhyay, M. Hoefer, J. Morschhäuser, R. Prasad, Expression of the CDR1 efflux pump in clinical Candida albicans isolates is controlled by a negative regulatory element, Biochemical and Biophysical Research Communications 332 (2005) 206–214, https://doi.org/10.1016/j.bbrc.2005.04.113.

[89] M. Gaur, N. Puri, R. Manoharlal, V. Rai, G. Mukhopadhayay, D. Choudhury, R. Prasad, MFS transportome of the human pathogenic yeast Candida albicans, BMC Genomics 9 (2008) 579, https://doi.org/10.1186/1471-2164-9-579.

[90] M.V. Keniya, E. Fleischer, A. Klinger, R.D. Cannon, B.C. Monk, Inhibitors of the Candida albicans major facilitator superfamily transporter Mdr1p responsible for fluconazole resistance, PLoS One 10 (2015) e0126350, https://doi.org/10.1371/journal.pone.0126350.

[91] Y. Zhou, M. Liao, C. Zhu, Y. Hu, T. Tong, X. Peng, M. Li, M. Feng, L. Cheng, B. Ren, X. Zhou, ERG3 and ERG11 genes are critical for the pathogenesis of Candida albicans during the oral mucosal infection, International Journal of Oral Science 10 (2018) 1–8, https://doi.org/10.1038/s41368-018-0013-2.

[92] C.M. Martel, J.E. Parker, O. Bader, M. Weig, U. Gross, A.G.S. Warrilow, N. Rolley, D.E. Kelly, S.L. Kelly, Identification and characterization of four azole-resistant erg3 mutants of Candida albicans, Antimicrobial

Agents and Chemotherapy 54 (2010) 4527–4533, https://doi.org/10.1128/AAC.00348-10.

[93] J.R. Blankenship, J. Heitman, Calcineurin is required for Candida albicans to survive calcium stress in serum, Infection and Immunity 73 (2005) 5767–5774, https://doi.org/10.1128/IAI.73.9.5767-5774.2005.

[94] H. Schwartz, B. Scroggins, A. Zuehlke, T. Kijima, K. Beebe, A. Mishra, L. Neckers, T. Prince, Combined HSP90 and kinase inhibitor therapy: insights from the cancer genome atlas, Cell Stress and Chaperones 20 (2015) 729–741, https://doi.org/10.1007/s12192-015-0604-1.

[95] P.K. Mukherjee, J. Chandra, D.M. Kuhn, M.A. Ghannoum, Mechanism of fluconazole resistance in Candida albicans biofilms: phase-specific role of efflux pumps and membrane sterols, Infection and Immunity 71 (2003) 4333–4340, https://doi.org/10.1128/IAI.71.8.4333-4340.2003.

[96] M.L. Villasmil, A.D. Barbosa, J.L. Cunningham, S. Siniossoglou, J.T.N. Jr, An Erg11 lanosterol 14-α-demethylase-Arv1 complex is required for Candida albicans virulence, PLoS One 15 (2020) e0235746, https://doi.org/10.1371/journal.pone.0235746.

[97] J.L. Song, J. Beth Harry, R.T. Eastman, B.G. Oliver, T.C. White, The Candida albicans lanosterol 14-α-Demethylase (ERG11) gene promoter is maximally induced after prolonged growth with antifungal drugs, Antimicrobial Agents and Chemotherapy 48 (2004) 1136–1144, https://doi.org/10.1128/AAC.48.4.1136-1144.2004.

[98] P. Vandeputte, S. Ferrari, A.T. Coste, Antifungal resistance and new strategies to control fungal infections, International Journal of Microbiology 2012 (2012), https://doi.org/10.1155/2012/713687.

[99] M. Baginski, K. Sternal, J. Czub, E. Borowski, Molecular modelling of membrane activity of amphotericin B, a polyene macrolide antifungal antibiotic, Acta Biochimica Polonica 52 (2005) 655–658.

[100] W.L. Chaffin, Candida albicans cell wall proteins, Microbiology and Molecular Biology Reviews: MMBR. 72 (2008) 495–544, https://doi.org/10.1128/MMBR.00032-07.

[101] D.S. Perlin, Resistance to echinocandin-class antifungal drugs, Drug Resistance Updates 10 (2007) 121–130, https://doi.org/10.1016/j.drup.2007.04.002.

[102] M.A. Pfaller, D.J. Diekema, Epidemiology of invasive candidiasis: a persistent public health problem, Clinical Microbiology Reviews 20 (2007) 133–163, https://doi.org/10.1128/CMR.00029-06.

[103] P.R. Juvvadi, F. Lamoth, W.J. Steinbach, Calcineurin as a multifunctional regulator: unraveling novel functions in fungal stress responses, hyphal growth, drug resistance, and pathogenesis, Fungal Biology Reviews 28 (2014) 56–69, https://doi.org/10.1016/j.fbr.2014.02.004.

[104] S.D. Singh, N. Robbins, A.K. Zaas, W.A. Schell, J.R. Perfect, L.E. Cowen, Hsp90 governs echinocandin resistance in the pathogenic yeast Candida albicans via calcineurin, PLoS Pathogens 5 (2009) e1000532, https://doi.org/10.1371/journal.ppat.1000532.

[105] W. Chang, J. Liu, M. Zhang, H. Shi, S. Zheng, X. Jin, Y. Gao, S. Wang, A. Ji, H. Lou, Efflux pump-mediated resistance to antifungal compounds can be prevented by conjugation with triphenylphosphonium cation, Nature Communications 9 (2018) 5102, https://doi.org/10.1038/s41467-018-07633-9.

[106] M. Zhang, F. Zhao, S. Wang, S. Lv, Y. Mou, C. Yao, Y. Zhou, F. Li, Molecular mechanism of azoles resistant Candida albicans in a patient with chronic mucocutaneous candidiasis, BMC Infectious Diseases 20 (2020) 126, https://doi.org/10.1186/s12879-020-4856-8.

CHAPTER 4

Multidrug transporters of fungal pathogen Candida

Introduction

Multidrug transporters are studied in detail in context of drug resistance and overexpression of efflux pumps is one of the major mechanisms of drug resistance in *Candida albicans*. There are two major classes of multidrug transporters namely; ABC (ATP binding cassette) transporter and MFS (Major facilitator superfamily) transporter. ABC (*CDR1*, *CDR2*) and MFS (*MDR1*) transporters work independently or in combination to impart drug resistance. These membrane proteins use various energy sources to effectively translocate compounds through cell membranes. ABC proteins are ATP-hydrolysis-dependent primary transporters. The MFS pumps are secondary transporters that cross the plasma membrane using the proton-motive force. Both types of transporters have distinct protein domains that confer substrate specificity: nucleotide binding domains (NBDs) in ABC pumps and transmembrane domains (TMDs) in both ABC and MFS pumps [1]. Candida cells contain two types of efflux pumps that are known to contribute to drug resistance: ATP binding cassette (ABC) transporters and major facilitators' superfamily (MFS) transporters and they are described below:

CONTENTS

Introduction ... 65
ABC transporters .. 65
MFS transporters .. 66
Regulation of MDR 66
PDR1 *and* PDR3 *are xenobiotic receptor in* S. cerevisiae ... 67
Xenobiotic regulation 68
Efflux pumps regulation 70
References 72

ABC transporters

ABC transporters are found in every cell of an organism to transport substances across the plasma membrane. However, some transporters have developed in such a way that they can now carry structurally unrelated substrates such as P-gp which enables a cancerous cell resistant to therapeutic drugs used. The most basic structure of ABC transporters consists of two NBDs and two TMDs. The TMDs span the membrane, typically six times, through putative-helices, and the NBDs are involved in ATP binding and hydrolysis. The NBDs and TMDs inside the pump polypeptide are arranged differently depending on the form of ABC protein [2].

MFS transporters

MFS transporters, like ABC transporters, are broad superfamilies of proteins with high sequence similarity that are present in plants, animals, bacteria, and fungi. DHA1 (drug: H antiporter 1; 12 TMS) and DHA2 (drug: H antiporter 2; 12 TMS) are two subfamilies of MFS transporters involved in drug efflux (14 TMS). CaMDR1 (also known as BENr) was the first MFS transporter gene discovered in a pathogenic fungus [3]. The ability of this gene to confer benomyl and methotrexate resistance on Saccharomyces cerevisiae led to its cloning. CaMDR1 overexpression has been found in both FLC resistant strains derived in vitro and azole resistant clinical isolates. DHA1 MFS transporters include CaMdr1p. TMS5 amino acid residues are essential for drug/Htransport, according to structural and functional studies of CaMdr1p. FLU1 is another DHA1 MFS gene found in Candida albicans [4]. It has also been reported that disruption of FLU1 in C. albicans had little effect on FLC susceptibility but made cells sensitive to mycophenolic acid, suggesting that it might be a pump substrate [5].

The MFS transporters operate through a proton gradient. CaMDR1 (C. albicans Multidrug Resistance) have been shown to be overexpressed, MDR1 (previously named BENr, associated with benomyl resistance in S. cerevisiae) gene deletion in resistant strains of C. albicans does not result in increased susceptibility to azoles.

Regulation of MDR

Multidrug resistance (MDR) is a significant side effect of treating opportunistic fungal infections in immunocompromised people, such as transplant recipients and cancer patients receiving cytotoxic chemotherapy. Enhanced understanding of the molecular mechanisms that regulate MDR in pathogenic fungi could aid in the creation of new therapies to combat these tenacious infections. MDR often is caused by fungal member's upregulating drug efflux pumps.

Pdr1p orthologs, for example, belong to the zinc-cluster transcription-factor. The molecular mechanisms, on the other hand, are poorly known. Pdr1p family members in *S. cerevisiae* and the human pathogen *C. glabrata* bind to structurally diverse drugs and xenobiotics, causing drug efflux pumps to be activated and MDR to be induced. Notably, this is mechanistically similar to the PXR nuclear receptor's regulation of MDR in vertebrates, showing an unlikely functional analogy between fungal and metazoan MDR regulators [6,7].

One of the important networks working in *S. cerevisiae* to counter drug treatment is called PDR (Pleiotropic Drug Resistance) and involves the two major players Pdr1p and Pdr3p defined by indispensable PDRE domain. *PDR5* and *SNQ2* were the first genes identified as target of Pdr1p and Pdr3p; both contain PDREs in their promoters. Thereafter *PDR10*, *PDR15*, and *YOR1* were also found to be target of these TFs as their transcription is strongly affected by Pdr1p and Pdr3p. Interestingly, there are non ABC proteins *HXT9* and *HXT11* belonging to the family of hexose transporter are reported to be regulated by these transcription factors [8]. The second family of transcription factors controlling drug transporter genes belongs to bZip protein family related to mammalian AP-1 transcription factors called as YAP family [9].

PDR1 and *PDR3* are xenobiotic receptor in *S. cerevisiae*

Chemically distinct drugs and xenobiotics, such as the antifungal ketoconazole, the translation inhibitor cycloheximide, and the classic PXR agonist rifampicin, can induce the expression of ATP-dependent drug efflux pumps (for example, PDR5) and other Pdr1p/Pdr3p target genes (for example, PDR16) in *S. cerevisiae* in a Pdr1p/Pdr3. The glucocorticoid receptor agonist dexamethasone, on the other hand, consistently failed to induce Pdr1p/Pdr3p-regulated genes. Since the mammalian nuclear receptor PXR regulates MDR gene expression by binding directly to xenobiotics, and since Pdr1p/Pdr3p have interesting functional similarities to PXR, it was investigated if Pdr1p and Pdr3p could also interact directly with xenobiotics to stimulate expression of their target genes (ketoconazole and cycloheximide) (Fig. 4.1).

An antifungal drug could bind to the Pdr1p/Pdr3p transcription factors' XBD domain, which binds to the promoter sequence of drug-efflux genes like PDR5. This would enable the Mediator complex's Gal11p subunit to bind to the drug–Pdr1/Pdr3p–DNA complex through its KIX domain. The Mediator complex would then promote RNA polymerase II (Pol II) recruitment and increased PDR5 transcription. PDR5 is a member of a gene family that codes for drug-efflux pumps including ABC transporters. These pumps enable the antifungal drug to be effluxed from the cell membrane, reducing the fungal cell's susceptibility to the toxic compound.

Xenobiotic regulation

In response to xenobiotics, *Candida glabrata* has a highly conserved Pdr1p ortholog that also controls drug efflux pumps. It was hypothesized that CgPdr1p could directly bind to azoles and other xenobiotics to promote gene expression and MDR in *C. glabrata* based on a few studies with Pdr1p/Pdr3p in fungus *S. cerevisiae*. CgPdr1p is necessary for *C. glabrata*'s intrinsically high azole resistance, and xenobiotics stimulate expression of the drug efflux pump gene CDR2 in a CgPdr1p-dependent manner. *C. glabrata* has two distinct genes that have major sequence similarities to the GAL11 gene in *S. cerevisiae*. CgGAL11A gene deletion significantly reduced xenobiotic-dependent activation of CgCDR2, close to CgPDR1 gene deletion, while CgGAL11B gene deletion had no effect on CgCDR2 expression [10].

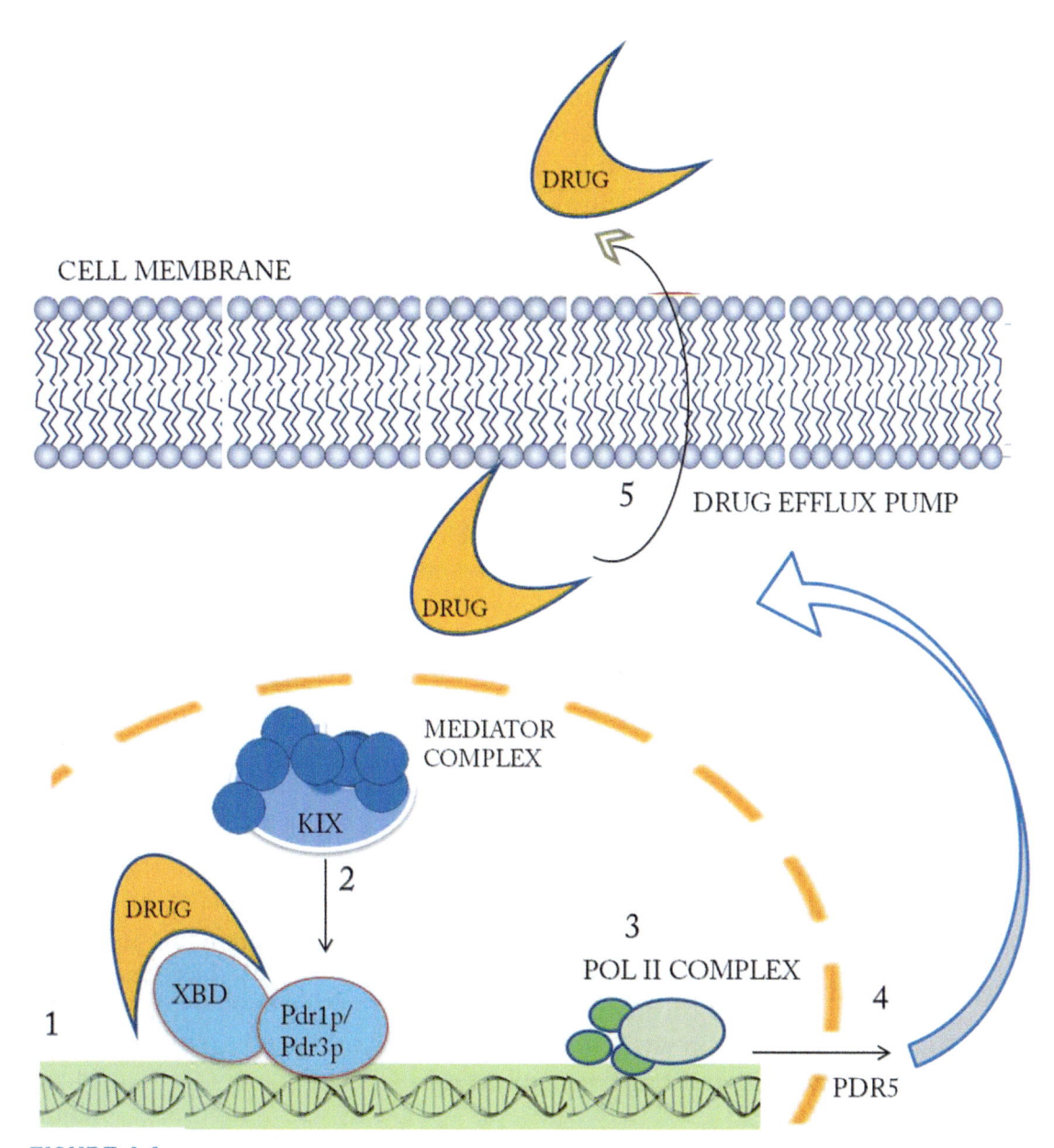

FIGURE 4.1
Events in pathogenic fungi on the exposure of drug.

Efflux pumps regulation

Cap1p was the first transcription factor to be linked to the regulation of multidrug resistance in *C. albicans* (*C. albicans* AP-1). Cap1p is a homolog of the *S. cerevisiae* bZip transcription factor Yap1p, which controls the transcription of genes encoding in oxidative stress resistance as well as members of the main facilitator and ABC transporter superfamilies. The ability of the CAP1 gene to impart fluconazole resistance in *S. cerevisiae* was used to identify it [11]. Cap1p-mediated upregulation of FLR1, which codes for a significant facilitator homologous to *C. albicans'* Mdr1p and confers tolerance to other toxic compounds, induced the fluconazole-resistant phenotype. Cap1 has lately been reported to bind to the MDR1 promoter in vivo [12].

In an in vitro generated MDR1 overexpressing *C. albicans* strain, however, deletion of CAP1 did not eliminate, but rather increased MDR1 expression, implying that the constitutive MDR1 overexpression in this strain was mediated by a separate, CAP1 independent mechanism. Cap1p, like Yap1p in *S. cerevisiae*, is essential for *C. albicans* oxidative stress resistance [11], and it was also found to be required for the induction of *MDR1* expression by H_2O_2 in a fluconazole-susceptible strain [13].

Mcm1p is another transcription factor implicated in the regulation of MDR1 expression. A putative Mcm1p binding site can be found in the MDRE/BRE of the MDR1 promoter [14]. As *MCM1* is an indispensable gene in the pathogen [15], in mcm1 null mutants, MDR1 expression cannot be observed [14]. Recent genome-wide location analyses confirmed that Mcm1p binds to the *MDR1* promoter in vivo but the requisite of Mcm1p for *MDR1* upregulation has not been recognized yet.

NDT80 was discovered in a screen for *C. albicans* genes that could cause a PCDR1-lacZ reporter fusion in the heterologous host *S. cerevisiae* [16]. Overexpression of CaNDT80 has improved resistance to fluconazole and ketoconazole, implying that it caused the expression of azole resistance genes. NDT80 expression was found to be induced in *C. albicans* in the presence of azoles, and deletion of NDT80 significantly reduced miconazole-induced CDR1 expression while having no effect on basal CDR1 expression levels. Although not as susceptible as a cdr1 deletion mutant, the *C. albicans* ndt80 deletion mutant was hypersusceptible to fluconazole and voriconazole [17].

As a result of the residual expression of Cdr1p in the tac1 deletion mutant, it was hypersusceptible to many antifungal drugs and other metabolic inhibitors to which Cdr1p and Cdr2p impart resistance, but not as prone as a mutant missing the efflux pumps themselves. Tac1 binds to the DRE and induces the CDR2 promoter in a DRE-dependent manner [18]. PDR16, which codes for the phosphatidylinositol transporter, was found to be coexpressed along with CDR1 and CDR2 in azole-resistant clinical isolates and responsive isolates on fluphenazine induction, suggesting that it is under Tac1p's power. The substitution of Asn to Asp at position 972 in TAC1 of drug-resistant strain 5674 was discovered. When this mutation is introduced into a wild-type strain, TAC1, PDR16, CDR1, and CDR2 are all inducible.

Mrr1p (multidrug resistance regulator) is a zinc cluster transcription protein that mediates MDR1 expression in *C. albicans*. MRR1 was discovered to be a gene that was upregulated in tandem with MDR1 in many fluconazole-resistant clinical *C. albicans* isolates, and its deletion eliminated MDR1 expression and drug resistance in these isolates [19]. In a drug-susceptible strain, MRR1 was also shown to be required for benomyl- and H_2O_2-induced MDR1 expression [20]. The same is true for *Candida dubliniensis* that produces an ortholog of MRR1 that regulates MDR1 overexpression in fluconazole-resistant strains [19].

MDR1 was first discovered as a resistance gene linked to the tetrahydrofolate reductase inhibitor methotrexate. FLU1 was first identified as a clone that conferred resistance to fluconazole [21]. Drug transporters play a role in antifungal drug resistance in non-albicans species like *C. glabrata*. In azole-resistant *C. glabrata* isolates, CgCDR1 and CgCDR2 are upregulated, rendering it a major nosocomial pathogen. It has also been suggested that two putative

ABC transporters, ABC1 and ABC2, can play a role in drug resistance in *C. krusei*. Fluconazole accumulation was reduced in azole-resistant *C. dubliniensis* isolates, leading to the discovery of two ABC transporters, CdCDR1 and CdCDR2 [22].

One of the most common mechanisms of resistance in fungi is drug efflux modification. CDR1 is the gene most commonly associated with energy-dependent drug efflux in fluconazole-resistant clinical isolates, and it is a homolog of *S. cerevisiae* PDR5, which encodes a multidrug efflux pump [23]. *C. albicans'* multidrug resistance phenotype has formerly been related to proteins encoded by the CDR1, CDR2, and MDR1 genes. These proteins act as membrane-bound efflux pumps, removing drugs from fungal cells [24]. The ABC family is responsible for transporting structurally unrelated drugs; this property has hampered therapeutic research because once tolerance is achieved for one type of agent, it is gained for all others translocated along with it, regardless of chemical class. The key proteins involved in major efflux mediated drug resistance are CaCdr1 and CaCdr2 [25].

References

[1] S.L. Panwar, R. Pasrija, R. Prasad, Membrane homoeostasis and multidrug resistance in yeast, Bioscience Reports 28 (2008) 217–228, https://doi.org/10.1042/BSR20080071.

[2] R.D. Cannon, A.R. Holmes, Learning the ABC of oral fungal drug resistance, Molecular Oral Microbiology (2015), https://doi.org/10.1111/omi.12109.

[3] R.D. Cannon, E. Lamping, A.R. Holmes, K. Niimi, P.V. Baret, M.V. Keniya, K. Tanabe, M. Niimi, A. Goffeau, B.C. Monk, Efflux-mediated antifungal drug resistance, Clinical Microbiology Reviews 22 (2009) 291–321, https://doi.org/10.1128/CMR.00051-08. Table of Contents.

[4] N. Sun, D. Li, W. Fonzi, X. Li, L. Zhang, R. Calderone, Multidrug-resistant transporter Mdr1p-mediated uptake of a novel antifungal compound, Antimicrobial Agents and Chemotherapy 57 (2013) 5931–5939, https://doi.org/10.1128/AAC.01504-13.

[5] R.D. Cannon, E. Lamping, A.R. Holmes, K. Niimi, K. Tanabe, M. Niimi, B.C. Monk, Candida albicans drug resistance another way to cope with stress, Microbiology 153 (2007) 3211–3217, https://doi.org/10.1099/mic.0.2007/010405-0.

[6] H.J. Lo, K.Y. Tseng, Y.Y. Kao, M.Y. Tsao, H.L. Lo, Y.L. Yang, Cph1p negatively regulates MDR1 involved in drug resistance in Candida albicans, International Journal of Antimicrobial Agents 45 (2015) 617–621, https://doi.org/10.1016/j.ijantimicag.2015.01.017.

[7] R. Li, R. Kumar, S. Tati, S. Puri, M. Edgerton, Candida albicans flu1-mediated efflux of salivary histatin 5 reduces its cytosolic concentration and fungicidal activity, Antimicrobial Agents and Chemotherapy 57 (2013) 1832–1839, https://doi.org/10.1128/AAC.02295-12.

[8] H.B. van den Hazel, H. Pichler, M.A. do Valle Matta, E. Leitner, A. Goffeau, G. Daum, PDR16 and PDR17, two homologous genes of Saccharomyces cerevisiae, affect lipid biosynthesis and resistance to multiple drugs, The Journal of Biological Chemistry 274 (1999) 1934–1941.

[9] I.S. A-correia, The yeast ABC transporter Pdr18 (ORF YNR070w) controls plasma, Biochemical Journal 202 (2011) 195–202, https://doi.org/10.1042/BJ20110876.

[10] J.P. Vermitsky, T.D. Edlind, Azole resistance in Candida glabrata: coordinate upregulation of multidrug transporters and evidence for a Pdr1-like transcription factor, Antimicrobial Agents and Chemotherapy 48 (2004) 3773–3781, https://doi.org/10.1128/AAC.48.10.3773-3781.2004.

[11] I.A.I. Hampe, J. Friedman, M. Edgerton, J. Morschhäuser, An acquired mechanism of antifungal drug resistance simultaneously enables Candida albicans to escape from intrinsic host defenses, PLoS Pathogens 13 (2017) e1006655, https://doi.org/10.1371/journal.ppat.1006655.

[12] S. Znaidi, K.S. Barker, S. Weber, A.-M. Alarco, T.T. Liu, G. Boucher, P.D. Rogers, M. Raymond, Identification of the Candida albicans Cap1p regulon, Eukaryotic Cell 8 (2009) 806–820, https://doi.org/10.1128/EC.00002-09.

[13] N. Dunkel, J. Blass, P.D. Rogers, J. Morschhäuser, Mutations in the multidrug resistance regulator MRR1, followed by loss of heterozygosity, are the main cause of MDR1 overexpression in fluconazole-resistant Candida albicans strains, Molecular Microbiology 69 (2008) 827–840, https://doi.org/10.1111/j.1365-2958.2008.06309.x.

[14] P.J. Riggle, C.A. Kumamoto, Transcriptional regulation of MDR1, encoding a drug efflux determinant, in fluconazole-resistant Candida albicans strains through an Mcm1p binding site, Eukaryotic Cell 5 (2006) 1957–1968, https://doi.org/10.1128/EC.00243-06.

[15] M. Rottmann, S. Dieter, H. Brunner, S. Rupp, A screen in Saccharomyces cerevisiae identified CaMCM1, an essential gene in Candida albicans crucial for morphogenesis, Molecular Microbiology 47 (2003) 943−959, https://doi.org/10.1046/j.1365-2958.2003.03358.x.

[16] A. Sellam, C. Askew, E. Epp, F. Tebbji, A. Mullick, M. Whiteway, A. Nantel, Role of transcription factor CaNdt80p in cell separation, hyphal growth, and virulence in Candida albicans, Eukaryotic Cell 9 (2010) 634−644, https://doi.org/10.1128/EC.00325-09.

[17] J. Branco, C. Martins-Cruz, L. Rodrigues, R.M. Silva, N. Araújo-Gomes, T. Gonçalves, I.M. Miranda, A.G. Rodrigues, The transcription factor Ndt80 is a repressor of Candida parapsilosis virulence attributes, Virulence 12 (2021) 601−614, https://doi.org/10.1080/21505594.2021.1878743.

[18] S.S. Krishnamurthy, R. Prasad, Membrane fluidity affects functions of Cdr1p, a multidrug ABC transporter of Candida albicans, FEMS Microbiology Letters 173 (1999) 475−481, https://doi.org/10.1016/S0378-1097(99)00113-5.

[19] J. Morschhäuser, K.S. Barker, T.T. Liu, J. Blaß-Warmuth, R. Homayouni, P.D. Rogers, The transcription factor Mrr1p controls expression of the MDR1 efflux pump and mediates multidrug resistance in Candida albicans, PLoS Pathogens 3 (2007) e164, https://doi.org/10.1371/journal.ppat.0030164.

[20] S. Schneider, J. Morschhäuser, Induction of Candida albicans drug resistance genes by hybrid zinc cluster transcription factors, Antimicrobial Agents and Chemotherapy 59 (2015) 558−569, https://doi.org/10.1128/AAC.04448-14.

[21] S. Schubert, P.D. Rogers, J. Morschhäuser, Gain-of-Function mutations in the transcription factor MRR1 are responsible for overexpression of the MDR1 efflux pump in fluconazole-resistant Candida dubliniensis strains, Antimicrobial Agents and Chemotherapy 52 (2008) 4274−4280, https://doi.org/10.1128/AAC.00740-08.

[22] S. Dogra, S. Krishnamurthy, V. Gupta, B.L. Dixit, Asymmetric distribution of phosphatidylethanolamine in C. albicans : possible mediation by CDR1 , a multidrug transporter belonging to ATP binding cassette (ABC), Superfamily 121 (1999) 111−121.

[23] K. Nakamura, M. Niimi, K. Niimi, R. Ann, J.E. Yates, A. Decottignies, C. Brian, A. Goffeau, R.D. Cannon, A.R. Holmes, Functional expression of Candida albicans drug efflux pump Cdr1p in a Saccharomyces cerevisiae strain deficient in membrane transporters functional expression of Candida albicans drug efflux pump Cdr1p in a Saccharomyces cerevisiae strain deficient in membrane transporters, Antimicrobial Agents and Chemotherapy 45 (2001) 3366−3374, https://doi.org/10.1128/AAC.45.12.3366.

[24] P. Vandeputte, S. Ferrari, A.T. Coste, Antifungal resistance and new strategies to control fungal infections, International Journal of Microbiology 2012 (2012), https://doi.org/10.1155/2012/713687.

[25] S. Shukla, T. Prasad, R. Prasad, Molecular Mechanism of Antifungal Resistance, 2006, pp. 1–15.

CHAPTER 5

Essential anticandidal targets

Fungal cell wall

The cell wall of fungi is an attractive target for antifungal agents because the structure does not exist in the host cell, so molecules that inhibit its synthesis may have low toxicity to humans. Just one class of antifungal drugs, the echinocandins, targets the fungal cell wall, despite the fact that a variety of archetypal antibiotics (e.g., penicillin) target the bacterial cell wall. The success of these drugs suggests that there may be other potential targets in the region [1]. *Candida albicans* can cause life-threatening systemic infections by invading deeper tissues. Mannoproteins with O-glycosylated oligosaccharide and N-glycosylated polysaccharide moieties make up the outermost layer of *Candida* cell wall. Both carbohydrate moieties play a role in virulence and host–fungal interactions. The N-glycosylated polysaccharide has a comblike structure with a 1,6-linked backbone moiety and an oligomannose side chain primarily made up of 1,2-, 1,3-, and 1,2-linked mannose residues with a few phosphate groups [2].

- Glucan
  The most important structural polysaccharide of the fungal cell wall is glucan, which accounts for 50%–60% of the dry weight of the structure [3]. The majority of glucan polymers (65%–90%) are made up of, α-1,3 and 1,3 linkage glucose units, but there are also glucans with β-1,6, β-1,4, and α-1,4 links [4]. The 1,3-D-glucan is the most significant structural component of the wall, and it is covalently connected to the other components of the structure. The 1,3-D-glucan is made by glucan synthases, which are a group of enzymes found in the plasma membrane. Cell growth is affected when one of these genes is disrupted, but cell death occurs when both are removed [5]. Similarly, studies

CONTENTS

Fungal cell wall 77
Fungal membrane components ... 78
Inhibition of heat shock protein 90 79
Inhibition of calcineurin signaling 81
Signaling pathways 81
Cell cycle control pathways 82
Membrane transporters .. 83

Other surface targets 87

Cross-talk 88

References 89

are being conducted with B-1, 6-glucan as a possible target. With chitin and B-1,3-glucan, it forms a wide network, with the first class of proteins linked to B-1,6-glucan. There are also new leads exploring processes essential for cell wall biosynthesis.

- Chitin
  The amount of chitin in the fungal wall varies depending on the fungus's morphological stage. It makes up 1%—2% of the dry weight of yeast cell walls, but it can be as high as 10%—20% in filamentous fungi. Chitin is made from n-acetylglucosamine by the enzyme chitin synthase, which deposits chitin polymers next to the cytoplasmic membrane in the extracellular space. The percentage of chitin in the hyphae wall of *Candida albicans* is three times thicker than that of the yeasts [6].
- Glycoproteins
  Proteins make up around 40% of the dry weight of the cell wall in fungi and 25% of the dry weight of the cell wall of filamentous fungi. Glycoproteins are formed when most proteins are linked to carbohydrates via O or N linkages [7]. Cell wall proteins play a variety of roles, including maintaining cellular structure, adhesion processes, cellular defense against various substances, molecule absorption, signal transmission, and the synthesis and reorganization of wall components [8].

Fungal membrane components

Ergosterol is a component of fungal cell membrane, serving the same function that cholesterol serves in animal cells. In fungi, the sterol biosynthetic pathway leads to the formation of Ergosterol. The Ergosterol pathway in fungal pathogens is an attractive antimicrobial target because it is unique from the major sterol (cholesterol) producing pathway in human. Target word is used only when particular pathway is unique to microorganism. Azole resistance in *C. albicans* is occurs primarily through *ERG 11* which encodes 14α-lanosterol demethylase (CYP51, also known as P45014 DM) involved in sterol biosynthesis. Azoles inhibit this step in the Ergosterol biosynthesis in fungi by binding to and inhibiting CYP51 [9]. Its inhibition leads to high levels of 14 methylated sterols, which causes disruption of membrane structures. Another enzyme $\Delta^{5,6}$ desaturase (*ERG 3*), has been shown to contribute to azole's resistance. A defect in *ERG 3* leads to accumulation of 14α methylfecosterol instead of 14 α-methylergosta-8, 24(28) —dien-3β, 6-α diol [10]. Accumulations of sufficient amount of 14α-methylfecosterol compensate for Ergosterol in the membranes and thus contribute to azole resistance in *C. albicans*.

The most abundant sterol in fungal cell membranes is ergosterol, a 5, 7-diene oxysterol that controls permeability and fluidity. Ergosterol is the target of the majority of clinically available antifungals due to its critical roles, unique structural properties, and specific biosynthetic steps. The products of the ERG11, ERG1 gene, and ERG2 gene are all affected by at present antifungal drugs that interfere with ergosterol synthesis (sterol C-8 isomerase; morpholines). The remaining genes involved directly or indirectly for ergosterol biosynthesis could be investigated as prospective antifungal targets [11].

Inhibition of heat shock protein 90

In *Candidia albicans*, Hsp90 is one of the most studied heat shock protein. Hsp90 is implicated in thermal stability, morphogenesis, cell cycle control, apoptosis, and drug resistance, according to numerous reports [12]. *C. albicans* has maintained the conserved protein Hsp90 throughout evolution, allowing it to adapt to thermal stress when colonizing warm-blooded animals or thermally buffered niches. Hsp90 is a protein that regulates cellular circuitry that is necessary for critical morphogenetic transitions from yeast to filament. Inhibiting the functions of Hsp90 genetically or pharmacologically inhibits the development, maturation, and dispersal of *C. albicans* biofilms in vitro [13].

Hsp104 expression rises in *C. albicans* cells after a brief exposure to elevated temperatures. Hsp104 acts as a prosurvival mediator in response to rising temperatures, implying that it plays an important role in thermotolerance [14]. Heat shock proteins have strong dependence on histone acetyltransferases (HATs) [15] and histone deacetylases (HDACs) as they play a key role in acetylation and eventually affect gene expression [16]. So the key to successfully targeting HDACs in the treatment of candidiasis is to establish stable and selective inhibitors that can differentiate pathogens from hosts.

Hsp70 is strongly conserved across a wide range of organisms, from bacteria to mammals. Because of the high degree of conservation of N-terminal domains, Hsp70 from different sources have similar biochemical properties; each has a high-affinity ATP-binding site and a peptide-binding site [17]. In *C. albicans*, the main members of the Hsp70 family are cell surface Ssa1 and Ssa2. It has been documented that compromising Hsp70 inhibits *C. albicans* phagocytosis in host cells and reduces antifungal drug resistance in *C. albicans*. As a result, if Hsp70 overcomes the safety barriers posed by its highly conserved structure among eukaryotes, it could be a potential antifungal target against candidiasis.

HSP60, one of the most well-known members of this protein family, has been described as an immunodominant target of autoantibodies and autoimmune T cells in both healthy people and people with inflammatory diseases including arthritis, type 1 diabetes, and atherosclerosis [18]. HSP60 has also been linked to the induction of regulatory T cells by interacting with the innate immune system [19]. HSP60 causes innate immune cells including macrophages and dendritic cells to release proinflammatory mediators, as we and others have discovered [20]. Fungal Hsp60 facilitates strong immunological reactions by acting as immunodominant antigens in humoral and cellular responses. The cross-reactivity of fungal and human Hsp60 could point to a correlation between infection and autoimmunity. Hsp60 has been shown to function as an immunogenic catalyst in the orchestration of *C. albicans*-related diseases under thermal stress in a number of studies [21]. These results may help researchers better understand the host–pathogen relationship to identify potential targets.

Inhibition of calcineurin signaling

Calcium signal transduction in fungi has received a lot of attention in recent years because of its importance in fungi survival. Calcineurin (CN), one of the calcium homeostasis regulators, has been identified as a virulence factor in filamentous fungi, and calcium channel proteins have been discovered to be responsible for filamentation of these pathogenic fungi [22]. In pathogenic fungi, calcineurin is a conserved calmodulin-dependent phosphatase that controls stress responses and is activated by the second messenger calcium [23]. *C. albicans* needs calcineurin to withstand cell membrane stress, cation stress, alkaline pH, and endoplasmic reticulum stress. Various receptors, transporters, and other proteins or enzymes in the calcium-calcineurin signaling system are closely linked to various physiological processes in *C. albicans*. Both CRZ1 and CNA1 are calcineurin pathway components that have been shown to modulate azole resistance in *C. albicans* [24].

Signaling pathways

In eukaryotic cells, mitogen activated protein (MAP) kinase signaling pathways control growth and stress adaptations such as thermal stress, apoptosis, and inflammation. Many extracellular stimuli trigger MAPK signaling pathways, which mediate signal transduction from the cell surface to the nucleus. In *C. albicans*, four MAPK signaling pathways have been discovered: the Mkc1 pathway, the Hog1 pathway [25], the Cek1 pathway [26], and the Cek2 pathway [27]. These signaling pathways are responsible for direct or indirect involvement in cell wall integrity; osmotic, oxidative, and other stress adaptations; mating; and starvation. In *C. albicans*, MAPK signaling pathways are closely linked to heat shock responses. MAPK's components—Cek1, Hog1, and Mkc1—are all Hsp90 client proteins. Hsp90 dysfunction inhibits cell wall biogenesis in *C. albicans* by impairing Cek1, Hog1, and Mkc1 activation. Thus, Hsp90 regulates long-term thermal adaptation through cell wall remodeling mediated by Mkc1, Hog1, and Cek1 [28].

The PKA complex, which consists of two catalytic subunits, Tpk1 and Tpk2, as well as a regulatory subunit, Bcy1, is activated by increased cAMP levels. It phosphorylates the downstream transcription factor Efg1, which activates a morphogenesis-regulating transcriptional program. *C. albicans* can invade several environmental niches in the host by modifying morphology between yeast and hyphal growth forms and modulating gene expression patterns. The PKA holoenzyme in *C. albicans* comprises two "inactivate" catalytic subunits (Tpk1 and Tpk2) that are each bound to a homodimer regulatory subunit (Bcy1) [29]. During Ras/cAMP/PKA signaling in *C. albicans*, activation of many proteins takes place that results in their respective virulence traits; Efg1 and Tec1 for cell adhesion [30]; Bcr1 and Rob 1 during biofilm formation [31]; Csc25, Cyr1, Srv2, and Tpk1/Tpk2 for filamentous growth, and white-to-opaque (W—O) switching [32].

The drug rapamycin targets TORs, which control cell development. Rapamycin (sirolimus) binds to the FKBP receptor, forming a rapamycin-FKBP complex that binds to the TOR protein and inhibits signal transduction [33]. Rapamycin inhibits filamentation in a variety of human and plant pathogens, which is a conserved mechanism of action among eukaryotes [34]. With several studies indicating that the TOR protein plays a globally conserved role in controlling growth and proliferation, targeting its critical functions should result in successful antifungal activity.

Cell cycle control pathways

The cell cycle is regulated by directional molecular events involving regulatory molecules such as cyclins and cyclin-dependent kinases (CDKs) [35]. Ccn1 and Cln3 are G1 cyclins that promote apical growth and bud elongation by interacting with Cdc28 during the *C. albicans* cell cycle [36]. G1 cyclins are degraded and replaced by G2 cyclins, Clb2 and Clb4, which bind to Cdc28 and suppress polarized growth, driving isotropic bud expansion, and thus modulating cyclin synthesis or degradation dramatically changes cell morphology [37]. CLB2 and CLB4, which are B cyclins, are involved in this operation, as is CLN3, which is an important G1 cyclin [38]. Regulators including cyclin-dependent kinases, checkpoint proteins, the proteasome, the heat shock protein Hsp90, and the heat shock transcription factor Hsf1 all play a role in morphogenesis, and their impact on the cell cycle and proteostasis are sometimes interconnected [39]. These findings show that several factors regulate morphogenesis in *C. albicans* through multiple elements of cell cycle control pathways. Modulation of these proteins thus appears to be a successful antifungal strategy for treating *C. albicans* infections.

Membrane transporters

Since cell transporters are involved in several drug resistance mechanisms, these surface targets could be studied in greater detail. MRP family includes 7 MRP genes, respectively, Other names used for MRP1 is ABCC1 and MRP, for MRP2 synonyms are ABCC2, cMOAT, and cMRP, for MRP3, it is ABCC3, MOAT-D, and cMOAT-2. MRP4 is also called as ABCC4 and MOAT-B. Other common names of MRP5 include ABCC5, MOAT-C, and pABC11, for MRP6 includes ABCC6, MOAT-E, MLP-1, and ARA†, and lastly MRP7 is also called as ABCC10 [40].

TMDs and NBDs are found in *C. albicans* ABC proteins; ten are full-size, three are full-size with an N-terminal TMD extension, five are half-size, and one is estimated to have two TMDs and one NBD, while the others do not have a TMD. Furthermore, the PDR subfamily includes seven full-size transporters, only two of which are confirmed to have multidrug transport capabilities (Tables 5.1 and 5.2) [40].

CDR2 mutants were found to be susceptible to azoles in previous experiments, and its overexpression conferred fluconazole tolerance. CDR1 mutants, on the other hand, showed an increased hypersensitivity [50]. CDR3, a PDR subclass protein that shares sequence homology with CDR1 and CDR 2, has been found to play no significant role in drug resistance. Cdr4 can play a role in osmotic and heavy metal stress responses. CDR 3 and 4 genes,

Table 5.1 List of transporter family proteins.

Protein family	Members	References
MDR (multidrug resistance)	Ste6 Atm1 Mdl I MDl II	[41]
PDR (pleiotropic drug resistance)	Pdr5 Snq2 Pdr10 Pdr12 Pdr15 Pdr11 Aus1 Adp1 Pdr18 YOL075C	[42]
MRP (multidrug resistance-associated protein)	Yor1 Ycf1 MRP1 MRP2 MRP3 MRP4 MRP5 MRP6 MRP7	[43]

on the other hand, encode flippase and translocate membrane phospholipids. The function and localization of CDR 5/ CDR 11 are still unknown [51]. Yor1 is a *C. albicans* ABC transporter that is found on the plasma membrane and is involved in aureobasidin A resistance. Other than CDR1 and CDR2, there has been no evidence of other ABC family members being involved in fluconazole resistance; among these, CDR 1 plays a relatively important role [52].

These are secondary active transporters that efflux substrates out of the cell by using the electrochemical gradient of protons across the plasma membrane. *C. albicans* has 95 putative MFS proteins grouped into 17 families, according to a genome-wide inventory [53]. Only CaMdr1p is known to extrude drugs out of all MFS

Table 5.2 List of ABC family proteins involved in *Candida albicans*.

ABC protein	Function	Subfamily	References
CaCdr1	Drug efflux, phospholipid translocation	PDR	[40]
CaCdr2	Drug efflux, phospholipid translocation	PDR	[16]
CaCdr3	Phospholipid translocation	PDR	[44]
CaCdr4	Phospholipid translocation	PDR	[45]
CaCdr5	Phospholipid translocation	PDR	[46]
CaHst6	Involved in mating of MTLa cells	MDR	[47]
CaMlt1	Vacuolar membrane transporter	MRP	[42]
CaYor1	Hyphal growth	PDR	[42]
CaYcf1	Vacuolar glutathione transporter	MRP	[42]
CaELF1	mRNA export protein	EF3	[48]
CaCEF3	Cytoplasmic translation elongation	EF3	[49]

proteins, and its overexpression has been related to azole resistance. FLU1 is yet another MFS gene (DHA1) whose absence has little impact on FLC, as predicted, but there is evidence that it is a pump substrate, as cells responsive to the immunosuppressant mycophenolic acid after disruption.

Upregulation of CDR and MDR follows two distinct pathways that are completely independent of one another. In comparison to MFS pumps, a lot of work has been done in the field of ABC expression and azole resistance. The development of new antifungal drugs and the prevention of drug resistance have aided in the understanding of the fungal pathogen's basic biology. Antifungal therapy faces a number of challenges, including developing drugs that target factors specific to fungi, which can be difficult given that fungi are eukaryotic and share several conserved biological pathways. Drug production could target genes that are critical for fungal survival. TAC1, for example, is a transcriptional activator of CDR genes that is found in the nucleus. TAC 1 mutants were found to have a lack of CDR1 and CDR2 upregulation. CAP1 is a transcription factor that inhibits MDR1 and has been linked to drug resistance and the oxidative stress response.

The FCR1 and FCR3 genes encode a transcription that belongs to the B-Zip family, but there is no evidence that they are involved in antifungal resistance. CDR gene expression is regulated by the Transcription Factor NDT80 gene. NDT80 was found to be essential for the expression of ERG genes such as ERG2, ERG25, ERG4, ERG24, ERG13, ERG9, ERG3, ERG10, ERG251, ERG1, ERG5, ERG6, ERG7, ERG11, and ERG12 [54]. Since NDT80 regulates sterol metabolism and drug resistance in *C. albicans*, it is an important component of this yeast's drug response [55].

Few ABC proteins are mainly found in organellar membranes, such as Mlt1, which is found in the vacuolar membrane and can confer metal resistance through intracellular sequestration in the vacuole. The yeast *Saccharomyces cerevisiae* was the first to record it. TOR stands for Target of Rapamycin, a group of proteins involved in the activation and regulation of protein synthesis. Elongation factors, EF-1 and EF-2, are involved in mammalian protein translation, but fungi need an additional factor called EF-3 for successful translation. Since it is not found in humans, this factor's uniqueness may be effectively exploited. ERG2, ERG3, ERG 5, ERG10, ERG 11, ERG 25, and NCP1 are *Candida albicans* genes involved in lipid, fatty acid, and sterol synthesis [56]. Azole antifungals are targets of the ERG 11 gene product [57]. C5 sterol desaturase is encoded by ERG3. Mutations in ERG 3 reduce ergosterol levels in the membrane and confer resistance to amphotericin. Proteins involved in cell wall maintenance are coded for by the genes PHR1, ECM21, ECM33, and FEN 12. Caspofungin affects the expression of these genes. The survival and pathogenicity of fungi include the maintenance of cell calcium homeostasis. Calcium homeostasis mechanisms and calcium signaling pathways have been linked to a variety of physiological processes in *C. albicans*, including stress responses, virulence, hyphal growth, and adhesion, according to researches [58]. Fungi can breach and invade host tissues thanks to secreted hydrolytic enzymes. Hydrolytic enzymes secreted recently have gotten a lot of attention as possible virulence factors in fungi. Protein degrading enzyme, lipid degrading enzymes, lipases, and phospholipases are the most well-known extracellular hydrolytic enzymes [59]. The absence or reduced expression of these hydrolytic enzymes has been contributing to *C. albicans* morphological transformation, colonization, cytotoxicity, and host penetration in a number of studies [60]. Other potential pathogenic mechanisms and targets in *C. albicans* that have not been thoroughly studied, such as the arachidonic acid metabolic pathway [61] and reactive oxygen species (ROS) homeostasis, are intriguing [62]. Prostaglandin production from arachidonic acid is needed for *C. albicans* growth and may be a major virulence factor in biofilm-associated infections, according to new research [63].

Other surface targets

Several research groups have stated that certain molecules interfere with glycosylphosphatidylinositol (GPI) biosynthesis. Membrane and cell wall homeostasis is dependent on GPI-anchored proteins. As previously mentioned, GPI proteins form a network with B-1,6-glucan. The elements that are peculiar to fungi are often studied in depth. Other elements, such as microtubule synthesis, signal transduction, and the cell cycle, may be used. Since they are not shared with the host, focusing on them increases target specificity.

Microtubules are polymers of the tubulin dimers and tubulin. Cell morphology and development are influenced by microtubule aggregation and disaggregation. Griseofulvin works by interacting with tubulin, a highly conserved protein in eukaryotes, to prevent microtubule aggregation. Nonetheless, there tend to be distinctions between mammalian and fungal tubulins; colchicine, for example, binds to mammalian tubulin preferentially.

Cross-talk

The host—pathogen mediated cross-talk inside *C. albicans* must be thoroughly understood in order to improve antifungal therapeutics. This intricate interplay can yield some potential goal leads since the infection process involves a large number of proteins and other factors. For starters, some general regulators of transcription include SUA71, TBP1, STP1, STP2P, STP3, STP4 [64]. Major players of metabolism control are also involved in cross-talk such as RGT1, TYE7, GAL4, MIG1 for carbohydrate metabolism, CBF1, GLN3, GCN4 for translation and amino acid metabolism, INO2, OPI1, CTF1 for lipid metabolism [65]. Lastly, some other factor for cell cycle and stress tolerance are SWI4, ASH1, CAT8, HAC1, CAS5, respectively [66]. NDT80 binds to the promoter regions of MFS drug transporters including MDR1 and FLU1. Multidrug resistance regulator 1 was the transcription factor involved in MDR1 upregulation in a clinical strain (MRR1). MRR1 inactivation resulted in the loss of MDR1 expression and increased fluconazole sensitivity in azole-resistant isolates. Scientists have successfully shown that Mrr1 controls MDR1 [67]. MRR1 has yet to be shown to bind directly to the MDR1 promoter, but it is possible that this transcription factor binds the BRE or MDRE regions directly or indirectly. The regulator of efflux pump 1 (REP1), which is part of the MDR1 family, can also regulate it [68]. MDR1 may still be upregulated in the presence of a medication despite the absence of both REP1 and MRR1. UPC2 is a transcriptional activator of ERG11, and upregulation of ERG 11 has been stated to fail in the absence of UPC2 [69].

Ada2 is an adaptor protein involved in many aspects of gene regulation. TAC1, MRR1, and UPC2 recruit it, which then regulate their own set of dependent genes [70]. There is cross-talk between the genes regulated by these major transcriptional activators and their target genes. TAC1 and UPC2 regulate CDR1, while CAP1 regulates both PDR16 and MDR1 [71]. This is a great illustration of genes controlling their subsets in various regulons. Only a few regulatory factors may target other factors that are connected to other transcriptional units. EFG1, a morphogenesis regulator, targets both CAP1 and NDT80. Despite the fact that the full range of interactions has yet to be investigated, the evidence so far indicates complex relationships.

Discovery of drugs with new targets is even more critical if they appear to adopt the exact pathway with the same target, allowing the pathogen to spawn resistance repeatedly. Since the paradigm of drug discovery has shifted from "one drug, one target" to "one drug, multiple targets" in the modern period, there is a pressing need for not only new leads, but also new targets and targeting approaches.

References

[1] N. Shibata, A. Suzuki, H. Kobayashi, Y. Okawa, Chemical structure of the cell-wall mannan of *Candida albicans* serotype A and its difference in yeast and hyphal forms, Biochemical Journal 404 (2007) 365–372, https://doi.org/10.1042/BJ20070081.

[2] A. Suzuki, Y. Takata, A. Oshie, A. Tezuka, N. Shibata, H. Kobayashi, Y. Okawa, S. Suzuki, Detection of beta-1,2-mannosyltransferase in *Candida albicans* cells, FEBS Letters 373 (1995) 275–279, https://doi.org/10.1016/0014-5793(95)01061-i.

[3] J.S. Piotrowski, H. Okada, F. Lu, S.C. Li, L. Hinchman, A. Ranjan, D.L. Smith, A.J. Higbee, A. Ulbrich, J.J. Coon, R. Deshpande, Y.V. Bukhman, S. McIlwain, I.M. Ong, C.L. Myers, C. Boone, R. Landick, J. Ralph, M. Kabbage, Y. Ohya, Plant-derived antifungal agent poacic acid targets beta-1,3-glucan, Proceedings of the National Academy of Sciences of the United States of America 112 (2015) E1490–E1497, https://doi.org/10.1073/pnas.1410400112.

[4] R.P. Hartland, G.W. Emerson, P.A. Sullivan, A secreted β-glucan-branching enzyme from *Candida albicans*, Proceedings of the Royal Society of London B: Biological Sciences (1991) 246. http://rspb.royalsocietypublishing.org/content/246/1316/155.long.

[5] R.C. Goldman, D.J. Frost, J.O. Capobianco, S. Kadam, R.R. Rasmussen, C. Abad-Zapatero, Antifungal drug targets: Candida secreted aspartyl protease and fungal wall beta-glucan synthesis, Infectious Agents and Disease 4 (1995) 228–247.

[6] W.L. Chaffin, *Candida albicans* cell wall proteins, Microbiology and Molecular Biology Reviews 72 (2008) 495–544, https://doi.org/10.1128/MMBR.00032-07.

[7] R. Garcia-Rubio, H.C. de Oliveira, J. Rivera, N. Trevijano-Contador, The fungal cell wall: *Candida, Cryptococcus,* and *Aspergillus* species, Frontiers in Microbiology 10 (2020), https://doi.org/10.3389/fmicb.2019.02993.

[8] M.V. Elorza, E. Valent, Molecular organization of the ell all of *Candida albicans* and its relation to pathogenicity, FEMS Yeast Research 6 (2006) 14−29, https://doi.org/10.1111/j.1567-1364.2005.00017.x.

[9] D. Van Booven, S. Marsh, H. McLeod, M.W. Carrillo, K. Sangkuhl, T.E. Klein, R.B. Altman, Cytochrome P450 2C9-CYP2C9, Pharmacogenetics and Genomics 20 (2010) 277−281, https://doi.org/10.1097/FPC.0b013e3283349e84.

[10] W.J. Hoekstra, E.P. Garvey, W.R. Moore, S.W. Rafferty, C.M. Yates, R.J. Schotzinger, Design and optimization of highly-selective fungal CYP51 inhibitors, Bioorganic & Medicinal Chemistry Letters 24 (2014) 3455−3458, https://doi.org/10.1016/j.bmcl.2014.05.068.

[11] L.-I. McCall, A. El Aroussi, J.Y. Choi, D.F. Vieira, G. De Muylder, J.B. Johnston, S. Chen, D. Kellar, J.L. Siqueira-Neto, W.R. Roush, L.M. Podust, J.H. McKerrow, Targeting ergosterol biosynthesis in *Leishmania donovani*: essentiality of sterol 14alpha-demethylase, PLOS Neglected Tropical Diseases 9 (2015), https://doi.org/10.1371/journal.pntd.0003588.

[12] L. Whitesell, N. Robbins, D.S. Huang, C.A. McLellan, T. Shekhar-Guturja, E.V. LeBlanc, C.S. Nation, R. Hui, A. Hutchinson, C. Collins, S. Chatterjee, R. Trilles, J.L. Xie, D.J. Krysan, S. Lindquist, J.A. Porco, U. Tatu, L.E. Brown, J. Pizarro, L.E. Cowen, Structural basis for species-selective targeting of Hsp90 in a pathogenic fungus, Nature Communications 10 (2019) 402, https://doi.org/10.1038/s41467-018-08248-w.

[13] T.R. O'Meara, N. Robbins, L.E. Cowen, The Hsp90 chaperone network modulates *Candida* virulence traits, Trends in Microbiology 25 (2017) 809−819, https://doi.org/10.1016/j.tim.2017.05.003.

[14] R.K. Chaudhary, J. Kardani, K. Singh, R. Banerjee, I. Roy, Deciphering the roles of trehalose and Hsp104 in the inhibition of aggregation of mutant huntingtin in a yeast model of Huntington's disease, NeuroMolecular Medicine 16 (2014) 280−291, https://doi.org/10.1007/s12017-013-8275-5.

[15] J. Soroka, S.K. Wandinger, N. Mäusbacher, T. Schreiber, K. Richter, H. Daub, J. Buchner, Conformational switching of the molecular chaperone Hsp90 via regulated phosphorylation, Molecular Cell 45 (2012) 517−528, https://doi.org/10.1016/j.molcel.2011.12.031.

[16] W.L. Smith, T.D. Edlind, Histone deacetylase inhibitors enhance *Candida albicans* sensitivity to azoles and related antifungals: correlation with reduction in CDR and ERG upregulation, Antimicrobial Agents and

Chemotherapy 46 (2002) 3532–3539, https://doi.org/10.1128/aac.46.11.3532-3539.2002.

[17] J.R. Glover, S. Lindquist, Hsp104, Hsp70, and Hsp40: a novel chaperone system that rescues previously aggregated proteins, Cell 94 (1998) 73–82, https://doi.org/10.1016/S0092-8674(00)81223-4.

[18] W. van Eden, R. van der Zee, B. Prakken, Heat-shock proteins induce T-cell regulation of chronic inflammation, Nature Reviews Immunology 5 (2005) 318–330, https://doi.org/10.1038/nri1593.

[19] S. Kamphuis, W. Kuis, W. de Jager, G. Teklenburg, M. Massa, G. Gordon, M. Boerhof, G.T. Rijkers, C.S. Uiterwaal, H.G. Otten, A. Sette, S. Albani, B.J. Prakken, Tolerogenic immune responses to novel T-cell epitopes from heat-shock protein 60 in juvenile idiopathic arthritis, The Lancet 366 (2005) 50–56, https://doi.org/10.1016/S0140-6736(05)66827-4.

[20] Q. Xu, G. Schett, H. Perschinka, M. Mayr, G. Egger, F. Oberhollenzer, J. Willeit, S. Kiechl, G. Wick, Serum soluble heat shock protein 60 is elevated in subjects with atherosclerosis in a general population, Circulation 102 (2000) 14–20, https://doi.org/10.1161/01.CIR.102.1.14.

[21] C. Habich, K. Kempe, F.J. Gomez, M. Lillicrap, H. Gaston, R. van der Zee, H. Kolb, V. Burkart, Heat shock protein 60: identification of specific epitopes for binding to primary macrophages, FEBS Letters 580 (2006) 115–120, https://doi.org/10.1016/j.febslet.2005.11.060.

[22] S.L. LaFayette, C. Collins, A.K. Zaas, W.A. Schell, M. Betancourt-Quiroz, A.A.L. Gunatilaka, J.R. Perfect, L.E. Cowen, PKC signaling regulates drug resistance of the fungal pathogen *Candida albicans* via circuitry comprised of Mkc1, calcineurin, and Hsp90, PLoS Pathogens 6 (2010) e1001069, https://doi.org/10.1371/journal.ppat.1001069.

[23] S. Liu, Y. Hou, W. Liu, C. Lu, W. Wang, S. Sun, Components of the calcium-calcineurin signaling pathway in fungal cells and their potential as antifungal targets, Eukaryot Cell 14 (2015) 324–334, https://doi.org/10.1128/EC.00271-14.

[24] X. Li, C. Yu, X. Huang, S. Sun, Synergistic effects and mechanisms of budesonide in combination with fluconazole against resistant *Candida albicans*, PLoS ONE 11 (2016) e0168936, https://doi.org/10.1371/journal.pone.0168936.

[25] I. Correia, R. Alonso-Monge, J. Pla, The Hog1 MAP kinase promotes the recovery from cell cycle arrest induced by hydrogen peroxide in *Candida albicans*, Frontiers in Microbiology 7 (2017), https://doi.org/10.3389/fmicb.2016.02133.

[26] A. Jha, A. Vimal, A. Bakht, A. Kumar, Inhibitors of CPH1-MAP kinase pathway: ascertaining potential ligands as multi-target drug candidate in *Candida albicans*, International Journal of Peptide Research and Therapeutics 25 (2019), https://doi.org/10.1007/s10989-018-9747-0.

[27] J. Chen, Q. Wang, J.-Y. Chen, CEK2, a novel MAPK from *Candida albicans* complement the mating defect of fus3/kss1 mutant, Sheng Wu Hua Xue Yu Sheng Wu Wu Li Xue Bao (Shanghai) 32 (2000) 299–304.

[28] I.D. Jacobsen, D. Wilson, B. Wächtler, S. Brunke, J.R. Naglik, B. Hube, *Candida albicans* dimorphism as a therapeutic target, Expert Review of Anti-infective Therapy 10 (2012) 85–93, https://doi.org/10.1586/eri.11.152.

[29] C.-J. Lin, Y.-L. Chen, Conserved and divergent functions of the cAMP/PKA signaling pathway in *Candida albicans* and *Candida tropicalis*, Journal of Fungi (Basel). 4 (2018), https://doi.org/10.3390/jof4020068.

[30] D.O. Inglis, G. Sherlock, Ras signaling gets fine-tuned: regulation of multiple pathogenic traits of *Candida albicans*, Eukaryot Cell 12 (2013) 1316–1325, https://doi.org/10.1128/EC.00094-13.

[31] G. Huang, Q. Huang, Y. Wei, Y. Wang, H. Du, Multiple roles and diverse regulation of the Ras/cAMP/protein kinase A pathway in Candida albicans, Molecular Microbiology 111 (2019) 6–16, https://doi.org/10.1111/mmi.14148.

[32] H. Si, A.D. Hernday, M.P. Hirakawa, A.D. Johnson, R.J. Bennett, *Candida albicans* white and opaque cells undergo distinct programs of filamentous growth, PLoS Pathogens 9 (2013), https://doi.org/10.1371/journal.ppat.1003210 e1003210–e1003210.

[33] M.C. Cruz, A.L. Goldstein, J. Blankenship, M. Del Poeta, J.R. Perfect, J.H. McCusker, Y.L. Bennani, M.E. Cardenas, J. Heitman, Rapamycin and less immunosuppressive analogs are toxic to *Candida albicans* and *Cryptococcus neoformans* via FKBP12-dependent inhibition of TOR, Antimicrobial Agents and Chemotherapy 45 (2001) 3162–3170, https://doi.org/10.1128/AAC.45.11.3162-3170.2001.

[34] N.-N. Liu, P.R. Flanagan, J. Zeng, N.M. Jani, M.E. Cardenas, G.P. Moran, J.R. Köhler, Phosphate is the third nutrient monitored by TOR in *Candida albicans* and provides a target for fungal-specific indirect TOR inhibition, Proceedings of the National Academy of Sciences of the United States of America 114 (2017) 6346–6351, https://doi.org/10.1073/pnas.1617799114.

[35] P.A. Pope, S. Bhaduri, P.M. Pryciak, Regulation of cyclin-substrate docking by a G1 arrest signaling pathway and the Cdk inhibitor Far1, Current Biology 24 (2014) 1390–1396, https://doi.org/10.1016/j.cub.2014.05.002.

[36] Y. Wang, CDKs and the yeast-hyphal decision, Current Opinion in Microbiology 12 (2009) 644–649, https://doi.org/10.1016/j.mib.2009.09.002.

[37] I. Correia, R. Alonso-Monge, J. Pla, MAPK cell-cycle regulation in *Saccharomyces cerevisiae* and *Candida albicans*, Future Microbiology 5 (2010) 1125–1141, https://doi.org/10.2217/fmb.10.72.

[38] J. Berman, Morphogenesis and cell cycle progression in *Candida albicans*, Current Opinion in Microbiology 9 (2006) 595–601, https://doi.org/10.1016/j.mib.2006.10.007.

[39] S. Hossain, E. Lash, A.O. Veri, L.E. Cowen, Functional connections between cell cycle and proteostasis in the regulation of *Candida albicans* morphogenesis, Cell Reports 34 (2021) 108781, https://doi.org/10.1016/j.celrep.2021.108781.

[40] R. Prasad, A. Banerjee, N.K. Khandelwal, S. Dhamgaye, The ABCs of *Candida albicans* multidrug transporter Cdr1, Eukaryotic Cell 14 (2015) 1154–1164, https://doi.org/10.1128/EC.00137-15.

[41] A. K Redhu, A.H. Shah, R. Prasad, MFS transporters of *Candida* species and their role in clinical drug resistance, FEMS Yeast Research 16 (2016), https://doi.org/10.1093/femsyr/fow043.

[42] E. Lamping, P.V. Baret, A.R. Holmes, B.C. Monk, A. Goffeau, R.D. Cannon, Fungal PDR transporters: phylogeny, topology, motifs and function, Fungal Genetics and Biology 47 (2010) 127, https://doi.org/10.1016/j.fgb.2009.10.007.

[43] A. Banerjee, J. Pata, S. Sharma, B.C. Monk, P. Falson, R. Prasad, Directed mutational strategies reveal drug binding and transport by the MDR transporters of *Candida albicans*, Journal of Fungi (Basel) 7 (2021), https://doi.org/10.3390/jof7020068.

[44] I. Balan, A.M. Alarco, M. Raymond, The *Candida albicans* CDR3 gene codes for an opaque-phase ABC transporter, Journal of Bacteriology 179 (1997) 7210–7218, https://doi.org/10.1128/jb.179.23.7210-7218.1997.

[45] S.J. Fox, B.T. Shelton, M.D. Kruppa, Characterization of genetic determinants that modulate *Candida albicans* filamentation in the presence of bacteria, PLOS ONE 8 (2013) e71939, https://doi.org/10.1371/journal.pone.0071939.

[46] N.Z.N. Nguyen, V.G. Tran, S. Lee, M. Kim, S.W. Kang, J. Kim, H.J. Kim, J.S. Lee, H.R. Cho, B. Kwon, CCR5-mediated recruitment of NK cells to the kidney is a critical step for host defense to systemic *Candida albicans* infection, Immune Network 20 (2020), https://doi.org/10.4110/in.2020.20.e49.

[47] H. Chibana, N. Oka, H. Nakayama, T. Aoyama, B.B. Magee, P.T. Magee, Y. Mikami, Sequence finishing and gene mapping for *Candida albicans* chromosome 7 and syntenic analysis against the *Saccharomyces cerevisiae* genome, Genetics 170 (2005) 1525–1537, https://doi.org/10.1534/genetics.104.034652.

[48] J. Sturtevant, R. Cihlar, R. Calderone, Disruption studies of a *Candida albicans* gene, ELF1: a member of the ATP-binding cassette family,

Microbiology (Reading) 144 (Pt 8) (1998) 2311–2321, https://doi.org/10.1099/00221287-144-8-2311.

[49] K.K. Myers, W.A. Fonzi, P.S. Sypherd, Isolation and sequence analysis of the gene for translation elongation factor 3 from *Candida albicans*, Nucleic Acids Research 20 (1992) 1705–1710, https://doi.org/10.1093/nar/20.7.1705.

[50] M.K. Rawal, A. Banerjee, A.H. Shah, M.F. Khan, S. Sen, A.K. Saxena, B.C. Monk, R.D. Cannon, R. Bhatnagar, A.K. Mondal, R. Prasad, Newly identified motifs in *Candida albicans* Cdr1 protein nucleotide binding domains are pleiotropic drug resistance subfamily-specific and functionally asymmetric, Scientific Reports 6 (2016) 27132, https://doi.org/10.1038/srep27132.

[51] R. Franz, S. Michel, J. Morschhäuser, A fourth gene from the *Candida albicans* CDR family of ABC transporters, Gene 220 (1998) 91–98, https://doi.org/10.1016/S0378-1119(98)00412-0.

[52] L.M. Chen, Y.H. Xu, C.L. Zhou, J. Zhao, C.Y. Li, R. Wang, Overexpression of CDR1 and CDR2 genes plays an important role in fluconazole resistance in *Candida albicans* with G487T and T916C mutations, Journal of International Medical 38 (2010) 536–545, https://doi.org/10.1177/147323001003800216.

[53] J. Shao, M. Zhang, T. Wang, Y. Li, C. Wang, The roles of CDR1, CDR2, and MDR1 in kaempferol-induced suppression with fluconazole-resistant *Candida albicans*, Pharmaceutical Biology 54 (2016) 984–992, https://doi.org/10.3109/13880209.2015.1091483.

[54] L.A. Vale-Silva, A.T. Coste, F. Ischer, J.E. Parker, S.L. Kelly, E. Pinto, D. Sanglard, Azole resistance by loss of function of the sterol $\Delta^{5,6}$-desaturase gene (ERG3) in *Candida albicans* does not necessarily decrease virulence, Antimicrobial Agents and Chemotherapy 56 (2012) 1960–1968, https://doi.org/10.1128/AAC.05720-11.

[55] I.V. Ene, S. Brunke, A.J.P. Brown, B. Hube, Metabolism in fungal pathogenesis, Cold Spring Harbor Perspectives in Medicine 4 (2014) a019695, https://doi.org/10.1101/cshperspect.a019695.

[56] S.A. Flowers, B. Colón, S.G. Whaley, M.A. Schuler, P.D. Rogers, Contribution of clinically derived mutations in ERG11 to azole resistance in *Candida albicans*, Antimicrobial Agents and Chemotherapy 59 (2015) 450–460, https://doi.org/10.1128/AAC.03470-14.

[57] M.-J. Xiang, J.-Y. Liu, P.-H. Ni, S. Wang, C. Shi, B. Wei, Y.-X. Ni, H.-L. Ge, Erg11 mutations associated with azole resistance in clinical isolates of *Candida albicans*, FEMS Yeast Research 13 (2013) 386–393, https://doi.org/10.1111/1567-1364.12042.

[58] Q. Yu, C. Xiao, K. Zhang, C. Jia, X. Ding, B. Zhang, Y. Wang, M. Li, The calcium channel blocker verapamil inhibits oxidative stress response in

Candida albicans, Mycopathologia 177 (2014) 167–177, https://doi.org/10.1007/s11046-014-9735-7.

[59] M. Schaller, C. Borelli, H.C. Korting, B. Hube, Hydrolytic enzymes as virulence factors of *Candida albicans*, Mycoses 48 (2005) 365–377, https://doi.org/10.1111/j.1439-0507.2005.01165.x.

[60] N. Pandey, M.K. Gupta, R. Tilak, Extracellular hydrolytic enzyme activities of the different *Candida* spp. isolated from the blood of the intensive care unit-admitted patients, Journal of Laboratory Physicians 10 (2018) 392–396, https://doi.org/10.4103/JLP.JLP_81_18.

[61] C. Chevtzoff, E.D. Yoboue, A. Galinier, L. Casteilla, B. Daignan-Fornier, M. Rigoulet, A. Devin, Reactive oxygen species-mediated regulation of mitochondrial biogenesis in the yeast *Saccharomyces cerevisiae*, Journal of Biological Chemistry 285 (2010) 1733–1742, https://doi.org/10.1074/jbc.M109.019570.

[62] T. Ren, H. Zhu, L. Tian, Q. Yu, M. Li, *Candida albicans* infection disturbs the redox homeostasis system and induces reactive oxygen species accumulation for epithelial cell death, FEMS Yeast Research 20 (2020), https://doi.org/10.1093/femsyr/foz081.

[63] J.R. Erb-Downward, M.C. Noverr, Characterization of prostaglandin E2 production by *Candida albicans*, Infection and Immunity 75 (2007) 3498–3505, https://doi.org/10.1128/IAI.00232-07.

[64] K. Min, A. Biermann, D.A. Hogan, J.B. Konopka, Genetic analysis of NDT80 family transcription factors in *Candida albicans* using new CRISPR-Cas9 approaches, MSphere 3 (2018), https://doi.org/10.1128/mSphere.00545-18.

[65] C. Sasse, R. Schillig, F. Dierolf, M. Weyler, S. Schneider, S. Mogavero, P.D. Rogers, J. Morschhäuser, The transcription factor Ndt80 does not contribute to Mrr1-, Tac1-, and Upc2-mediated fluconazole resistance in *Candida albicans*, PLoS One 6 (2011) e25623, https://doi.org/10.1371/journal.pone.0025623.

[66] I. Nocedal, E. Mancera, A.D. Johnson, Gene regulatory network plasticity predates a switch in function of a conserved transcription regulator, ELife 6 (2017) e23250, https://doi.org/10.7554/eLife.23250.

[67] R. Prasad, A. Singh, Lipids of *Candida albicans* and their role in multidrug resistance, Current Genetics 59 (2013) 243–250, https://doi.org/10.1007/s00294-013-0402-1.

[68] C.M. Hull, J.E. Parker, O. Bader, M. Weig, U. Gross, A.G.S. Warrilow, D.E. Kelly, S.L. Kelly, Facultative sterol uptake in an ergosterol-deficient clinical isolate of *Candida glabrata* harboring a missense mutation in ERG11 and exhibiting cross-resistance to azoles and amphotericin B. https://doi.org/10.1128/AAC.06253-11, 2012.

[69] M. Gaur, D. Choudhury, R. Prasad, Complete inventory of ABC proteins in human pathogenic yeast, *Candida albicans*, Microbial Physiology 9 (2005) 3−15, https://doi.org/10.1159/000088141.

[70] S.-J. Yu, Y.-L. Chang, Y.-L. Chen, Deletion of ADA2 increases antifungal drug susceptibility and virulence in *Candida glabrata*, Antimicrobial Agents and Chemotherapy 62 (2018), https://doi.org/10.1128/AAC.01924-17.

[71] I. Kounatidis, L. Ames, R. Mistry, H. Ho, K. Haynes, P. Ligoxygakis, A host-pathogen interaction screen identifies ada2 as a mediator of *Candida glabrata* defenses against reactive oxygen species, G3: Genes, Genomes, Genetics 8 (2018) 1637−1647, https://doi.org/10.1534/g3.118.200182.

CHAPTER 6

Stages of anticandidal agent development

Introduction

In India, the advancement of new drug discovery has the potential to boost the country's overall healthcare system. The introduction of new medicines would aid in the alleviation of disease burdens since clinical trials performed in India would provide greater access to novel medicines. The expansion of the local pharmaceutical research and development sector can be a key driver of economic growth, resulting in higher income levels and better access to healthcare. Drug discovery and development is the process that brings a compound from scratch identification as a lead to a commercially available drug. During the course of drug development numerous strategies have been taken up which could be further subcategorized as, in vivo, in vitro, and in silico in their true forms. The advantages of in vitro experiments are that they provide quick results, are relatively inexpensive, and can be used to assess precise modes of action. The disadvantage of these experiments is that they lack the homeostatic processes and pathways found in animals [1]. In vitro experiments clearly show the effects of a substance in a regulated environment outside of a living organism, confirming the drug's legitimacy. However, in vivo experiments are best suited for confirming the results of an experiment on a living subject, as well as the involved mechanisms of action. In vivo trials, as opposed to in vitro tests, are more likely to produce significant findings with definitive and deducible insights into the essence of the drug and its effect on the host. In comparison, in silico studies are a relatively new form of investigation and study. Implementing in silico methods has led to major findings [2]. While in silico and in vitro models will continually be

CONTENTS

Introduction 97
Drug target discovery 99
Screening 99
HIT identification 100
Lead optimization 100
Experimentation .101
Characterization of product 101
Formulation, delivery, packaging development 102

Anticandidal Therapeutics. https://doi.org/10.1016/B978-0-443-18744-5.00003-4
Copyright © 2023 Elsevier Inc. All rights reserved.

CHAPTER 6: Stages of anticandidal agent development

Pharmacokinetics and drug disposition 102

Preclinical testing and IND application 102

Acute studies.. 103

Repeated dose studies 103

Generic toxicity studies 104

Reproductive toxicity studies 104

Carcinogenicity studies 104

Toxicokinetic studies 104

Bioanalytical testing 105

Clinical trials .. 105

Postmarketing surveillance 106

References 107

developed and refined, in vivo preclinical safety models remain the gold standard for assessing human risk [3]. Any drug development process must proceed through several stages in order to produce a product that is safe, efficacious, and has passed all regulatory requirements. The events involved in discovering a lead includes screening of natural compounds or synthesis of synthetic ones. This entire process of introducing a novel therapy from ground level to market usually takes 10–16 years (Fig. 6.1).

To find chemical compounds that can be produced and sold, a range of methods is used. According to the current state of chemical and biological sciences needed for pharmaceutical production, each new drug approved for human use must undergo laboratory screening of 5000–10,000 chemical compounds. Around 250 of the 5000–10,000 compounds screened will go

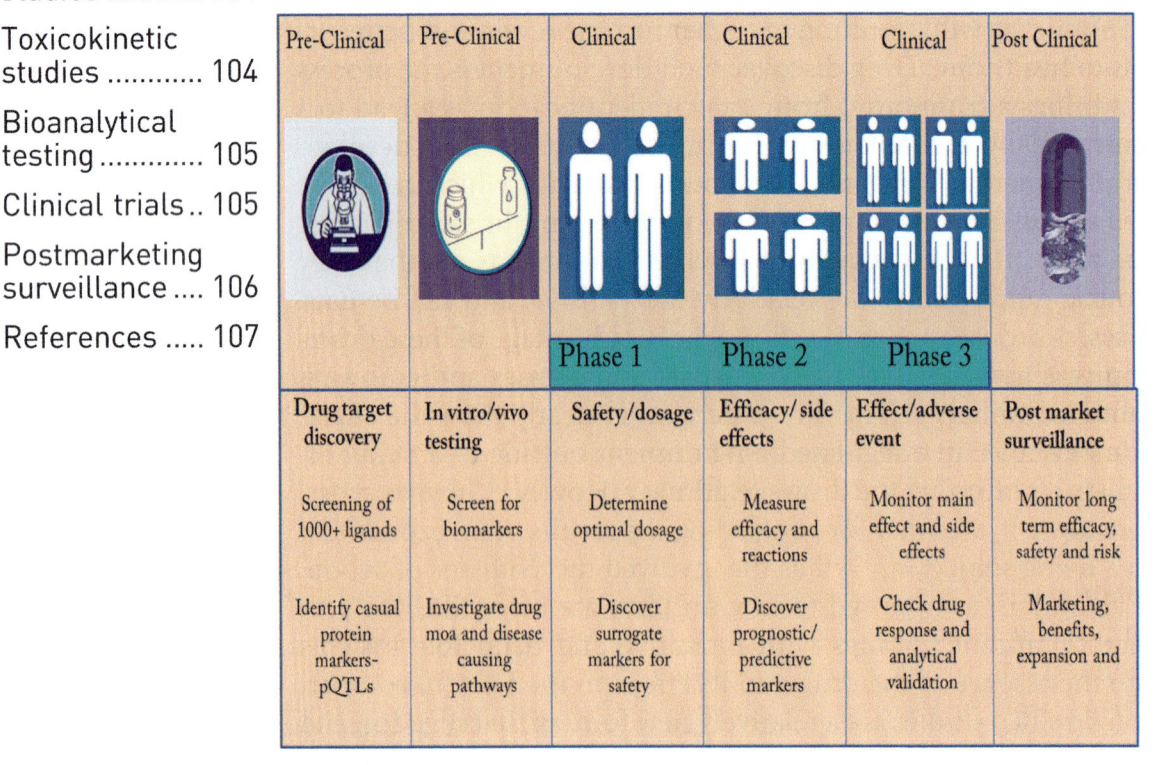

FIGURE 6.1

Various stages and phases of antifungal development.

into preclinical testing, and 5 will go into clinical testing. The entire process of developing a drug, from discovery to marketing, can take 10–15 years [4]. This segment explains how the pharmaceutical industry discovers and develops new medicines.

Drug target discovery

In academic and pharmaceutical/biotech research laboratories, drug candidates are evaluated for their association with the drug's target. Up to 10,000 molecules are subjected to a rigorous screening procedure for each potential drug *Candida*te, which may include functional genomics and/or proteomics, as well as other screening methods. Scientists normally look for activity against the disease disorder for which the medication is being developed after verifying interaction with the drug target. After careful consideration, one or more lead compounds are chosen [5]. When developing a "good" drug, it is important to provide a thorough understanding of the clinical spectrum of a disease as well as the precise role that a target plays in that disease. If a therapeutic agent—whether a small molecule or a biopharmaceutical—may alter the function of a biological target, it is said to be "druggable." Drug targets with "universal" beneficial properties can be identified in a number of ways, including scouring published scientific literature, checking available databases, or using "practical" approaches including target deconvolution and target discovery. If a goal has been determined, it is validated to ensure that it is suitable for pharmaceutical production before launching a screening campaign to find "hits."

Screening

For drug research and production, screening chemical compounds for possible pharmacological effects is a critical step. Almost every chemical and pharmaceutical company in the world has a library of chemical compounds that have been synthesized over a long period of time. Many different chemicals have been extracted from natural resources like plants, animals, and microorganisms in the past, and they are used to develop as antiinfectious agents [6–8]. University chemists have access to a wide range of chemical compounds [9]. Hundreds of thousands of new compounds have also been discovered thanks to automated, high-throughput combinatorial chemistry methods. Whether some of these millions of compounds possess the properties that will enable them to be developed into drugs is to be tested and identified.

HIT identification

To find a "hit" compound, a variety of screening methods may be used. A compound that interacts with the object of interest is referred to as a "strike." High-throughput screening, phenotypic screening, virtual screening, fragment-based screening, and structure-based design are some of the techniques that can be used to find "hits." Since phenotypic screens are focused on analyzing a shift in phenotype to determine whether a compound exerts a desired impact, the particular drug target cannot be immediately apparent. While it is not a regulatory necessity to know the target of a drug as long as it exhibits good safety and efficacy properties, the target underlying the observed phenotypic shift can be known later.

Lead optimization

At this point, the primary goal is to refine a few of the most promising "hits" in order to build more potent and selective candidates with "optimized" pharmacokinetic properties. Researchers are working to augment the affinity of the "initial hits" to the biological target by many orders of magnitude. Off-target interactions are also important to remember at this point because they can have negative consequences, so enhancing the molecule's selectivity against other biological targets should be investigated and discussed.

Identification of a clinical candidate is achieved from many successful "high-quality" leads. In addition to the above characteristics, you can think about potential manufacturing suitability and scale-up, market feasibility, and cost-effectiveness, as these will have a significant impact on the drug's long-term performance [10,11].

Experimentation

The discovery of new drugs using mouse, rat, and dog models is an example of in vivo preclinical research. In vitro testing is carried out in a laboratory. Animal cells or tissues from a nonliving animal are used in ex vivo research. Finding successful cancer treatment agents, measuring tissue properties, and practical modeling for new surgical techniques are all examples of ex vivo research assays. A cell is often used as the basis for small explant cultures in an ex vivo assay, which provides a complex, regulated, and sterile environment.

Intestinal, topical, membrane, intravenous, and oral are some of the drug delivery methods. Drug delivery systems are used to distribute new drugs in a managed or tailored manner. Physiological barriers in the body of an animal or a human can prevent drugs from reaching their intended target or from releasing when they should. The aim is to keep the drug successful while preventing it from interfering with healthy tissues. Intramuscular (IM), intraperitoneal (IP), and subcutaneous (SC) drug delivery are all examples of parenteral drug delivery (SC) [12]. It's often used on unconscious patients because it prevents difficult-to-cross epithelial barriers. One of the quickest drug delivery absorption methods is parenteral or intravenous injection. IV injection guarantees that all of the drugs are absorbed into the bloodstream, making it more powerful than IM, SC, or LP membrane approaches [11].

Characterization of product

When a drug lead molecule shows potential as a therapeutic, it has to be identified and characterized, which involves determining the physiochemical properties such as lead's size, shape, strengths and limitations, function, toxicity, bioactivity, and bioavailability.

Pharmacology studies at an early stage may help identify the compound's underlying mechanism of action [13].

Formulation, delivery, packaging development

Drug makers must create a formulation that guarantees the correct drug delivery parameters. At this stage in the drug development process, it's important to start thinking about clinical trials. The composition and distribution of a drug can be fine-tuned indefinitely, even after it has received final approval. Scientists assess the drug's stability in terms of its composition as well as all of the factors that affect storage and shipping, such as heat, light, and time. The formulation must be both potent and sterile, as well as effective (nontoxic) [14].

Pharmacokinetics and drug disposition

Formulation scientists may benefit from pharmacokinetic (PK) or ADME (Absorption/Distribution/Metabolism/Excretion) studies [15]. ADME stands for Absorption, Distribution, Disposition, Metabolism, and Excretion, and it is a PK method for determining how a new drug affects the body. Each effect in ADME is defined mathematically. Proof of Principle (PoP) studies is preclinical and early safety research studies have been successful. In drug research and development programs, the terms *Proof of Concept* (PoC) and *Proof of Principle* (PoP) are almost interchangeable. Following the completion of successful PoP/PoC tests, the program is advanced to Phase II dosage studies. AUC (area under the curve), Cmax (maximum drug concentration in blood), and Tmax are all parameters obtained from PK studies (time at which Cmax is reached). The predictive power of animal models is then tested by comparing data from early stage clinical trials with data from animal PK studies [16].

Preclinical testing and IND application

Preclinical testing is used to gather critical evidence about a lead molecule's effectiveness and safety before it is put to the test on humans [17]. In most cases, both in vitro and in vivo models are used to demonstrate its biological effect. APIs (active pharmaceutical ingredients) are biologically active ingredients in a drug Candidate that have a therapeutic effect. The API or APIs, as well as excipients, make up all medications. (Excipients are inactive ingredients that help the drug enter the human body.) HP APIs (High Potency Active Pharmaceutical Ingredients) are molecules that work at far lower doses than normal APIs. They are categorized by toxicity, pharmacological efficacy, and occupational exposure limits (OELs), and they are used in multistep drug growth. Regulatory agencies such as the FDA and the MHRA need preclinical trials until submitting an investigational new drug application (IND), which is expected to move on to clinical development [18].

Preclinical research examines the formulated drug product's bioactivity, safety, and efficacy. Many regulatory agencies scrutinize this research because it is crucial to a drug's ultimate success. Clinical trial plans and an Investigational New Drug (IND) application are planned during the initial phases of the development process. Studies conducted in the preclinical phase should be planned to facilitate subsequent clinical studies. The following are the primary phases of preclinical toxicology testing:

Acute studies

Acute toxicity assays analyze the effects of one or more doses given over a 24-h span. The aim is to figure out how much is too much and watch for clinical signs of toxicity. At least two mammalian species are typically tested. Acute toxicity data is used to help assess doses for repeated dose trials in animals and Phase I human studies.

Repeated dose studies

Repeated dose trials are classified as subacute, subchronic, or chronic, depending on how long they last. The period should be based on the length of the clinical trial for the new drug. In most cases, two species are needed.

Generic toxicity studies

These tests determine whether or not a drug compound is mutagenic or carcinogenic. The Ames test (conducted in bacteria) is used to detect genetic changes. The Mouse Micronucleus Test, for example, uses mammalian cells to assess DNA damage. The Chromosomal Aberration Test and other procedures look for chromosomal injury.

Reproductive toxicity studies

The effects of the drug on fertility are examined in Segment I reproductive tox studies. Effects on embryonic and postnatal development are discovered in Segment II and III studies. Before a medication may be given to women of childbearing age, reproductive toxicity trials must be performed.

Carcinogenicity studies

Only drugs designed to treat chronic or ongoing disorders need carcinogenicity studies. They take time and money, and they must be scheduled early in the preclinical testing phase.

Toxicokinetic studies

These are usually designed in the same way as PK/ADME studies, but with much higher dose levels. They investigate the effects of toxic drug doses and aid in determining the clinical margin of safety. There are several FDA and ICH recommendations that go into great detail about the various forms of preclinical toxicology studies and why they should be conducted in relation to IND, NDA, and BLA filings.

Bioanalytical testing

Much of the other processes in the drug development phase are supported by bioanalytical laboratory work. Bioanalytical work is essential for proper molecule characterization, assay production, developing optimal cell culture or fermentation methods, evaluating process yields, and ensuring quality assurance and quality control during the development process. It is also essential for preclinical toxicology and pharmacology research, as well as clinical trials.

Clinical trials

Clinical trials are designed to answer specific research questions regarding a medication that is currently being developed. A study protocol, which is a manual that specifies how the clinical trial will be carried out, must be followed. It outlines the key research objectives, study design, and methodological considerations to ensure the safety of participants and the quality of the data collected during the review [13]. Clinical trials are classified into three groups or stages based on their goals:

> Phase I Clinical Development (Human Pharmacology)—Unless the FDA places a hold on the study, a biopharmaceutical company will start a small-scale Phase I clinical trial a month after filing an IND. In most Phase I studies, healthy volunteers are used to determine pharmacological activity and tolerance. These studies include initial single-dose experiments, dose escalation studies, and quick repeated-dose studies [19].
> Phase II Clinical Development (Therapeutic Exploratory)—Phase II clinical trials are small-scale studies conducted in groups of 100–250 participants to determine a drug's potential efficacy and side-effect profile [20].
> Phase III Clinical Development (Therapeutic Confirmatory)—Phase III trials are large-scale clinical trials that examine the safety and efficacy of a large number of people. Although phase III trials are underway, provisions

are made for filing the Biologics License Application (BLA) or the New Drug Application (NDA) [21].

Globally, the review procedure for marketing authorization is known as a New Drug Application (NDA). This procedure is known as a Marketing Authorization Application in the European Union and other countries around the world (MAA). The scientific assessment of the NDA or MAA is the responsibility of the regulatory authority. The application's aim is to provide enough information to the regulator—gathered during preclinical and clinical trials—for them to decide whether the drug is safe and efficient, whether the drug's therapeutic benefits outweigh the risks, and whether the drug's manufacturing methods and quality assurance measures are adequate [22].

Once the new drug has been designed for maximum effectiveness and protection, as well as the outcomes of clinical trials, it is sent to the FDA for a comprehensive review. The FDA reviews and accepts or rejects the drug application submitted by the drug production company at this time [23].

Postmarketing surveillance

The word "post-marketing safety surveillance" refers to the monitoring of a drug after it has been approved and placed on the market. Its aim is to assess a drug's long-term safety and effectiveness, as well as possible "real-world" formulation issues and usage for unapproved or "off-label" conditions (e.g., use in an age group or at a dosage outside of that advised in the product label).

After a drug has been approved, phase IV trials are carried out. And after a new medication has been approved; the government should continue to track its safety in case any side effects occur when it is used in a wider population, interactions with other medications that were not evaluated in a premarketing clinical study, as well as its side effects [24].

Thousands of people took part in the case. The volunteers will be diagnosed with the illness or disorder for which the medication has been approved.

A Phase IV analysis is designed to gather more knowledge about the long-term dangers and benefits of taking a drug now that it is more commonly used [25]. The "real-world" data can also be used to assess if the drug has potential for further development, such as:

1. To see if the medication may be used for other uses or age ranges.
2. To devise a different method of administration.

In India, the advancement of new drug discovery has the potential to boost the country's overall healthcare system. The introduction of new medicines would aid in the alleviation of disease burdens since clinical trials performed in India would provide greater access to novel medicines. The expansion of the local pharmaceutical research and development sector can be a key driver of economic growth, resulting in higher income levels and better access to healthcare.

References

[1] P.K. Mishra, G.V. Raghuram, A. Bhargava, A. Ahirwar, R. Samarth, R. Upadhyaya, S.K. Jain, N. Pathak, In vitro and in vivo evaluation of the anticarcinogenic and cancer chemopreventive potential of a flavonoid-rich fraction from a traditional Indian herb *Selaginella bryopteris*, British Journal of Nutrition 106 (2011) 1154–1168, https://doi.org/10.1017/S0007114511001498.

[2] T. Kaur, A. Madgulkar, M. Bhalekar, K. Asgaonkar, Molecular docking in formulation and development, Current Drug Discovery Technologies 16 (2019) 30–39, https://doi.org/10.2174/1570163815666180219112421.

[3] X. Liu, D. Shi, S. Zhou, H. Liu, H. Liu, X. Yao, Molecular dynamics simulations and novel drug discovery, Expert Opinion on Drug Discovery 13 (2018) 23–37, https://doi.org/10.1080/17460441.2018.1403419.

[4] J.P. Hughes, S. Rees, S.B. Kalindjian, K.L. Philpott, Principles of early drug discovery, British Journal of Pharmacology 162 (2011) 1239–1249, https://doi.org/10.1111/j.1476-5381.2010.01127.x.

[5] J.-J. Lu, W. Pan, Y.-J. Hu, Y.-T. Wang, Multi-target drugs: the trend of drug research and development, PLoS One 7 (2012) e40262, https://doi.org/10.1371/journal.pone.0040262.

[6] N. Singh, A. Kumar, P. Gupta, R. Maurya, A. Dube, Evaluation of antileishmanial potential of Tinospora sinensis against experimental visceral leishmaniasis, Parasitology Research 102 (2008) 561−565.

[7] A. Vimal, A. Kumar, Spices chemoconstituents as persuasive inhibitor of *S. typhimurium* virulent protein L-asparaginase, Letters in Drug Design & Discovery 14 (12) (2017) 1433−1454.

[8] J.S. Prusty, A. Kumar, Coumarins: antifungal effectiveness and future therapeutic scope, Molecular Divers 24 (2020) 1367−1383.

[9] E. Jenwitheesuk, J.A. Horst, K.L. Rivas, W.C. Van Voorhis, R. Samudrala, S.M. Hammer, Novel paradigms for drug discovery: computational multitarget screening, Trends in Pharmacological Sciences 29 (2008) 62−71, https://doi.org/10.1016/j.tips.2007.11.007.

[10] J.K. Oh, R. Drumright, D.J. Siegwart, K. Matyjaszewski, The development of microgels/nanogels for drug delivery applications, Progress in Polymer Science (Oxford) 33 (2008) 448−477, https://doi.org/10.1016/j.progpolymsci.2008.01.002.

[11] A.V. Singh, Biopolymers in drug delivery: a review, Pharmacologyonline 1 (2011) 666−674.

[12] M. Feher, J.M. Schmidt, Property distributions: differences between drugs, natural products, and molecules from combinatorial chemistry (2002). https://doi.org/10.1021/CI0200467.

[13] E.S.D. Ashley, R. Lewis, J.S. Lewis, C. Martin, D. Andes, Pharmacology of systemic antifungal agents, Clinical Infectious Diseases 43 (2006) S28−S39, https://doi.org/10.1086/504492.

[14] T. Brodniewicz, G. Grynkiewicz, Preclinical drug development, Acta Poloniae Pharmaceutica 67 (2010) 578−585.

[15] T.D.Y. Chung, D.B. Terry, L.H. Smith, In vitro and in vivo assessment of ADME and PK properties during lead selection and lead optimization—guidelines, benchmarks and rules of thumb, Eli Lilly & Company and the National Center for Advancing Translational Sciences (2004). http://www.ncbi.nlm.nih.gov/pubmed/26561695.

[16] S. Ekins, J. Rose, In silico ADME/Tox: the state of the art, Journal of Molecular Graphics and Modelling 20 (2002) 305−309, https://doi.org/10.1016/S1093-3263(01)00127-9.

[17] K. Singh, S. Mehta, The clinical development process for a novel preventive vaccine: an overview, Journal of Postgraduate Medicine 62 (2016) 4−11, https://doi.org/10.4103/0022-3859.173187.

[18] F. on N. and N.S. Disorders, B. on H.S. Policy, I. of Medicine, Drug Development Challenges, National Academies Press (US), 2014. https://www.ncbi.nlm.nih.gov/books/NBK195047/ (Accessed 23 April 2021).

[19] J.K. Aronson, What is a clinical trial? British Journal of Clinical Pharmacology 58 (2004) 1–3, https://doi.org/10.1111/j.1365-2125.2004.02184.x.

[20] C.A. Umscheid, D.J. Margolis, C.E. Grossman, Key concepts of clinical trials: a narrative review, Postgraduate Medicine 123 (2011) 194–204, https://doi.org/10.3810/pgm.2011.09.2475.

[21] R. DerSimonian, N. Laird, Meta-analysis in clinical trials, Controlled Clinical Trials 7 (1986) 177–188, https://doi.org/10.1016/0197-2456(86)90046-2.

[22] R. Brinkmann, Post-marketing surveillance of psychotropic drugs, in: S.G. Dahl, L.F. Gram (Eds.), Clinical Pharmacology in Psychiatry, Springer, Berlin, Heidelberg, 1989, pp. 311–316, https://doi.org/10.1007/978-3-642-74430-3_36.

[23] R.L. Bell, E. O'Brian Smith, Clinical trials in post-marketing surveillance of drugs, Controlled Clinical Trials 3 (1982) 61–68, https://doi.org/10.1016/0197-2456(82)90019-8.

[24] N. Raj, S. Fernandes, N.R. Charyulu, A. Dubey, R. GS, S. Hebbar, Postmarket surveillance: a review on key aspects and measures on the effective functioning in the context of the United Kingdom and Canada, Therapeutic Advances in Drug Safety 10 (2019), https://doi.org/10.1177/2042098619865413, 2042098619865413.

[25] B.K. Chen, Y.T. Yang, Post-marketing surveillance of prescription drug safety: past, present, and future, Journal of Legal Medicine 34 (2013) 193–213, https://doi.org/10.1080/01947648.2013.800797.

CHAPTER 7

Recent trends and progress in antifungal translational research

Introduction

Recent studies have provided new information about the cellular and organic specific mechanistic basis of the host's immune response to fungi, receptors, and related pathways, and how changes in these pathways increase the susceptibility of fungi to fungal infections [1]. The data of various drugs have suggested that there are several causes of resistance developed and if we take a look on clinical and experimental studies on individual drugs currently used in treatment regime, we would get more detailed insight on translational improvement and development [2]. For instance, Walsh et al. have laid a foundation of improved treatment and new approaches for echinocandins as they are fighting not just nosocomial infections but biofilms on Central venous catheter CVC [3]. Here, micafungin function has been constantly updated for its stronger efficacy against not just biofilm but also planktonic *Candida* species [4]. In another study, recent models of catheter-related biofilm formation have been resolved with systemic therapy of micafungin. Since biofilms are major virulence factor of *Candidal* infection, these studies can be quite helpful in future. Another hurdle faced in fungal therapeutics is the need to constantly administer drugs intravenously. Petratiene et al. have created a strategy for extending the dose interval. Several researches have reported that eradication of infection with higher single doses could be used in patients of disseminated candidiasis [5].

Another approach proven to be extremely promising is the development of fungal vaccine [6]. Although the effectiveness of single antigen to be used as a "pan-fungal" vaccine is still in question,

CONTENTS

Introduction . 111

Translational research and strategy in fungal infection 112

References .. 116

specific preventive or adjunctive therapy is quite beneficial [7]. The production of effective vaccines to fight fungal infections has been aided by constant researches of host defense and pathogenic mechanisms. As a result, scientists have focused their efforts on developing robust, long-lasting, and safe effective agents, especially those that could be useful in endemic infections or chronic or superimposed infections in intensive care units [8]. Adjuvants and vaccine formulations are being improved in order to induce stronger defensive responses and to exploit immune response mechanisms that aren't compromised [9].

Despite the fact that only a few human clinical trials have been completed, an increasing number of antifungal vaccine candidates are being studied in preclinical studies [10]. This may be due to a renewed interest in the possible use of vaccines to mitigate antifungal medication use and, as a result, drug resistance and toxicity by replacing or combining them with chemotherapy. Invasive candidiasis research has made the most progress in the field of fungal vaccines, and two successful vaccines are currently in clinical trials. The first, containing the rAls3p-N antigen, is in Phase IIa and protects immunized hosts from fungal adhesion and invasion in disseminated and oropharyngeal candidiasis [11]. T cells, through neutrophil recruitment and a speck of antibodies, have been found to play a big role in this vaccine's safety [12]. A virosome-based vaccine containing Secreted Aspartyl Proteinases (Sap2 antigen/truncated recombinant Sap2) antigen is the second candidal vaccine in clinical trials. The truncated form is immunogenic, stable, and nontoxic [13]. It has been reported in a study by Kadosh et al. that systemic and complete mucosal protective immunity is caused by the Sap2 vaccine when given intramuscularly or intravaginally [14].

Translational research and strategy in fungal infection

As the use of antifungals increases in both the clinical and agricultural industries, the number of encounters with antifungal-resistant isolates will almost certainly increase as well. In order to maintain our existing selection of antifungal agents, antifungal stewardship would need to become a new focus for antimicrobial stewardship initiatives. Rapid diagnostics can aid stewardship efforts by reducing the time it takes to determine whether or not an antifungal agent is appropriate for a given patient.

The most striking feature of an invasive fungal infection is immunocompromised nature of patients. Since invasive fungal infections are most common in patients with weakened immune systems, many attempts have been made to put our understanding of defensive antifungal immunity into clinical practice [15]. To strengthen the further therapeutic backbone, several approaches based on antifungal immunology have been trending in the last few years [16].

It has been clearly reported that the number of circulating neutrophils in the body is directly proportional to the risk for catching fungal infections. To mitigate this, several studies have been reported recently that focus on transfusion therapy of granulocytes. Due to lesser data on this matter, the effect of this transfusion cannot be determined with confidence. Not only are these effective but they are also essentially the need of the hour for bench to bedside.

Moreover neutrophil transfusion might have certain adverse effects as well but the amount of risk mitigation will definitely overpower the limitations in the long run. Other than granulocytes, several other cells have also been used as transfusion therapy such as natural killer cells (NK), dendritic cells (DC), and cytokines. Natural killer cells are a type of immune effector that uses a perforin-dependent pathway to actively bind and destroy fungal pathogen. The receptor that regulates this mechanism is unknown, as is its possible function in disease [17]. Not only do NK cells kill *Candida* but they are also known to inhibit *Cryptococcus, Aspergillus, Rhizopus,* and *Paracoccidioides* [18]. Other than NK cells, RNA transfected dendritic cells are reported to be used for additional protection in murine transfection models [19]. This sudden increase in neutrophils count is actively reported to enhance the immune system of human host [20]. The understanding of the immunopathology of pathogenic fungi has contributed to the development of potential immunotherapy strategies. Increased phagocyte count, activation of innate host defense mechanisms in phagocytes and DC, and stimulation of antigen-specific immunity are among these techniques [21].

Among other strategies, cytokine transfusion or cytokines therapy is a very vast field that definitely will help in translating the benefits of its research in antifungal therapeutics. There are several cytokines that play a key role in protection and activation of immune system in human host like interleukin, interferons, and TNF [22].

For cytokines infusion, mainly colony stimulating factor and Interferon Gamma (IFN-ϒ) have been reported for adoptive immunotherapy [21].

While IFN-ϒ has been known as connecting link between innate and adaptive immunity, it has also been reported to protect patients with neutrophil deficiency a disease called as granulomatous disease [23]. In colony-stimulating factor, usage of infusion therapy to augment the immune response of human host has been effectively reported in past few years [24]. Patient suffering from hematopoietic transplantation have been infused with GM-CSF, which has resulted in decreased mortality rate up to 600 folds [25]. In another study conducted by Kelleher et al. they have reported that the usage of GM-CSF and a synergistic combinatorial response of both colony-stimulating factors have successfully alleviated mortality and morbidity [26].

Pattern recognition receptors (PRRs) have been very significant part of human immune system. PRR are molecules that recognize compounds present in pathogens on a regular basis. *Candida* and *Aspergillus* both have been reported to costimulate wide range of PRRs, like Dectin-1 which is required for fungal immunity [27].

In one study, Curdlan which is an agonistic molecule of dectin 1 has been suggested to aid in clearing fungal keratitis caused by *Aspergillus*, as it enhances the inflammatory response induced by the fungal pathogen in human epithelial cells [28]. Inspiration can be derived from its success against *Aspergillus* infection and researches against *Candida* can be initiated for the same. Similarly, there are many other researches that have been fruitful against *Aspergillus*; for instance adoptive therapy. In this approach, adoptive transfer of Aspergillus specific CD4 and CD8 cells have been reported to play essential role in antifungal immunity in animal models [29]. Although this therapy is still in its initial stages, the adoptive transfer of CD cells can be applied against *Candida*-based infections as well [22]. Although these researches have been reported with effective results, but for establishing them in *Candidal* therapeutic regime, a large-scale systematic clinical trial is necessary.

As we already know that due to prolonged use of drugs, resistance toward these agents has been significantly increasing. But the problem isn't limited to just the failure of current drugs; the entire process of novel drug development is delayed. To control this problem, further efforts in translational research is warranted. Although virulence factors have been increasing with time; it is needed of researchers to put in more efforts to establish a complete control over infection [19]. In recent years, as new strategies are under progress, researchers should also understand the need to master the most basic level of any regulatory pathway to develop an effective antifungal. If we overcome the unmet needs and the challenges to conduct a successful clinical trial rapidly, then it could shorten the period for drug development. Before translating the health benefits of any drug lead its detailed analysis is required not just in terms of trials but also in terms of efficacy and toxicity. Once we have established the efficacy and is ready for applicability, then we can successfully translate the benefits of that particular drug for human usage.

References

[1] M. Bruno, S. Kersten, J.M. Bain, M. Jaeger, D. Rosati, M.D. Kruppa, D.W. Lowman, P.J. Rice, B. Graves, Z. Ma, Y.N. Jiao, A. Chowdhary, G. Renieris, F.L. van de Veerdonk, B.-J. Kullberg, E.J. Giamarellos-Bourboulis, A. Hoischen, N.A.R. Gow, A.J.P. Brown, J.F. Meis, D.L. Williams, M.G. Netea, Transcriptional and functional insights into the host immune response against the emerging fungal pathogen Candida auris, Nature Microbiology 5 (2020) 1516–1531, https://doi.org/10.1038/s41564-020-0780-3.

[2] D.M. MacCallum, Hosting infection: experimental models to assay Candida virulence, International Journal of Microbiology 2012 (2011) e363764, https://doi.org/10.1155/2012/363764.

[3] T.J. Walsh, N. Azie, D.R. Andes, Development of new strategies for echinocandins: progress in translational research, Clinical Infectious Diseases 61 (2015) S601–S603, https://doi.org/10.1093/cid/civ676.

[4] S. Yamazaki, F. Nakamura, A. Yoshimi, M. Ichikawa, Y. Nannya, M. Kurokawa, Safety of high-dose micafungin for patients with hematological diseases, Leukemia & Lymphoma 55 (2014) 2572–2576, https://doi.org/10.3109/10428194.2014.885514.

[5] R. Petraitiene, V. Petraitis, W.W. Hope, T.J. Walsh, Intermittent dosing of micafungin is effective for treatment of experimental disseminated candidiasis in persistently neutropenic rabbits, Clinical Infectious Diseases 61 (2015) S643–S651, https://doi.org/10.1093/cid/civ817.

[6] A. Vecchiarelli, E. Pericolini, E. Gabrielli, D. Pietrella, New approaches in the development of a vaccine for mucosal candidiasis: progress and challenges, Frontiers in Microbiology 3 (2012), https://doi.org/10.3389/fmicb.2012.00294.

[7] T. Gebrehiwet, G. Gebremichael, Development of vaccination against fungal disease: a review article, International Journal of Tropical Diseases 1 (2018), https://doi.org/10.23937/ijtd-2017/1710005.

[8] A.M. Rauseo, A. Coler-Reilly, L. Larson, A. Spec, Hope on the horizon: novel fungal treatments in development, Open Forum Infectious Diseases 7 (2020), https://doi.org/10.1093/ofid/ofaa016.

[9] S. Nami, R. Mohammadi, M. Vakili, K. Khezripour, H. Mirzaei, H. Morovati, Fungal vaccines, mechanism of actions and immunology: a comprehensive review, Biomedicine & Pharmacotherapy 109 (2019) 333–344, https://doi.org/10.1016/j.biopha.2018.10.075.

[10] L.V.N. Oliveira, R. Wang, C.A. Specht, S.M. Levitz, Vaccines for human fungal diseases: close but still a long way to go, NPJ Vaccines 6 (2021) 33, https://doi.org/10.1038/s41541-021-00294-8.

[11] B.J. Spellberg, A.S. Ibrahim, V. Avanesian, Y. Fu, C. Myers, Q.T. Phan, S.G. Filler, M.R. Yeaman, J.E. Edwards Jr., Efficacy of the anti-Candida rAls3p-N or rAls1p-N vaccines against disseminated and mucosal candidiasis, The Journal of Infectious Diseases 194 (2006) 256–260, https://doi.org/10.1086/504691.

[12] S.A. Krumm, K.J. Doores, Targeting glycans on human pathogens for vaccine design, in: L. Hangartner, D.R. Burton (Eds.), Vaccination Strategies against Highly Variable Pathogens, Springer International Publishing, Cham, 2018, pp. 129–163, https://doi.org/10.1007/82_2018_103.

[13] M. Shukla, S. Rohatgi, Vaccination with secreted Aspartyl proteinase 2 protein from *Candida parapsilosis* can enhance survival of mice during *C. tropicalis*-mediated systemic candidiasis, Infection and Immunity 88 (2020), https://doi.org/10.1128/IAI.00312-20.

[14] D. Kadosh, Control of *Candida albicans* morphology and pathogenicity by post-transcriptional mechanisms, Cellular and Molecular Life Sciences 73 (2016) 4265–4278, https://doi.org/10.1007/s00018-016-2294-y.

[15] M. Seif, A. Häder, J. Löffler, O. Kurzai, From bench to bedside—translational approaches in anti-fungal immunology, Current Opinion in Microbiology 58 (2020) 153–159, https://doi.org/10.1016/j.mib.2020.10.004.

[16] R. Dinser, A. Grgic, Y.-J. Kim, M. Pfreundschuh, J. Schubert, Successful treatment of disseminated aspergillosis with the combination of voriconazole, caspofungin, granulocyte transfusions, and surgery followed by allogeneic blood stem cell transplantation in a patient with primary failure of an autologous stem cell graft, European Journal of Haematology 74 (2005) 438–441, https://doi.org/10.1111/j.1600-0609.2004.00384.x.

[17] H. Ogbomo, C.H. Mody, Granule-dependent natural killer cell cytotoxicity to fungal pathogens, Frontiers in Immunology 7 (2017), https://doi.org/10.3389/fimmu.2016.00692.

[18] S.S. Li, S.K. Kyei, M. Timm-McCann, H. Ogbomo, G.J. Jones, M. Shi, R.F. Xiang, P. Oykhman, S.M. Huston, A. Islam, M.J. Gill, S.M. Robbins, C.H. Mody, The NK receptor NKp30 mediates direct fungal recognition and killing and is diminished in NK cells from HIV-infected patients, Cell Host & Microbe 14 (2013) 387–397, https://doi.org/10.1016/j.chom.2013.09.007.

[19] L. Romani, Immunity to fungal infections, Nature Reviews Immunology 11 (2011) 275–288, https://doi.org/10.1038/nri2939.

[20] S. Bozza, K. Perruccio, C. Montagnoli, R. Gaziano, S. Bellocchio, E. Burchielli, G. Nkwanyuo, L. Pitzurra, A. Velardi, L. Romani, A dendritic cell vaccine against invasive aspergillosis in allogeneic hematopoietic transplantation, Blood 102 (2003) 3807–3814, https://doi.org/10.1182/blood-2003-03-0748.

[21] B.H. Segal, J. Kwon-Chung, T.J. Walsh, B.S. Klein, M. Battiwalla, N.G. Almyroudis, S.M. Holland, L. Romani, Immunotherapy for fungal infections, Clinical Infectious Diseases 42 (2006) 507—515, https://doi.org/10.1086/499811.

[22] A. Papadopoulou, P. Kaloyannidis, E. Yannaki, C.R. Cruz, Adoptive transfer of Aspergillus-specific T cells as a novel anti-fungal therapy for hematopoietic stem cell transplant recipients: progress and challenges, Critical Reviews in Oncology/Hematology 98 (2016) 62—72, https://doi.org/10.1016/j.critrevonc.2015.10.005.

[23] E.L. Assendorp, M.S. Gresnigt, E.G.G. Sprenkeler, J.F. Meis, N. Dors, J.W.M. van der Linden, S.S.V. Henriet, Adjunctive interferon-γ immunotherapy in a pediatric case of *Aspergillus terreus* infection, European Journal of Clinical Microbiology & Infectious Diseases 37 (2018) 1915—1922, https://doi.org/10.1007/s10096-018-3325-4.

[24] D. Armstrong-James, I.A. Teo, S. Shrivastava, M.A. Petrou, D. Taube, A. Dorling, S. Shaunak, Exogenous interferon-γ immunotherapy for invasive fungal infections in kidney transplant patients, American Journal of Transplantation 10 (2010) 1796—1803, https://doi.org/10.1111/j.1600-6143.2010.03094.x.

[25] J.N. Jarvis, G. Meintjes, K. Rebe, G.N. Williams, T. Bicanic, A. Williams, C. Schutz, L.-G. Bekker, R. Wood, T.S. Harrison, Adjunctive interferon-γ immunotherapy for the treatment of HIV-associated cryptococcal meningitis: a randomized controlled trial, AIDS 26 (2012) 1105—1113, https://doi.org/10.1097/QAD.0b013e3283536a93.

[26] P. Kelleher, A. Goodsall, A. Mulgirigama, H. Kunst, D.C. Henderson, R. Wilson, A. Newman-Taylor, M. Levin, Interferon-γ therapy in two patients with progressive chronic pulmonary aspergillosis, European Respiratory Journal 27 (2006) 1307—1310, https://doi.org/10.1183/09031936.06.00021705.

[27] M. da Glória Sousa, D.M. Reid, E. Schweighoffer, V. Tybulewicz, J. Ruland, J. Langhorne, S. Yamasaki, P.R. Taylor, S.R. Almeida, G.D. Brown, Restoration of pattern recognition receptor costimulation to treat chromoblastomycosis, a chronic fungal infection of the skin, Cell Host & Microbe 9 (2011) 436—443, https://doi.org/10.1016/j.chom.2011.04.005.

[28] W. Belda, P.R. Criado, L.F.D. Passero, Successful treatment of chromoblastomycosis caused by *Fonsecaea pedrosoi* using imiquimod, The Journal of Dermatology 47 (2020) 409—412, https://doi.org/10.1111/1346-8138.15225.

[29] Z. Sun, P. Zhu, L. Li, Z. Wan, Z. Zhao, R. Li, Adoptive immunity mediated by HLA-A*0201 restricted Asp f16 peptides-specific CD8+ T cells against Aspergillus fumigatus infection, European Journal of Clinical Microbiology & Infectious diseases 31 (2012) 3089—3096, https://doi.org/10.1007/s10096-012-1670-2.

CHAPTER 8

Clinical status of anticandidal therapeutic agents

Introduction

Clinical trials serve three purposes: (1) to decide if a medication is successful in the treatment (or prophylaxis) of a disease or disorder; (2) to evaluate the drug's risks and safety profile; and (3) to determine the overall risk-benefit ratio for patients who will be treated with the drug for a specific condition. A proper clarification of these elements is unlikely to be obtained in a single trial, but rather through a series of investigations, with each stage of the investigation building on the previous phase and trials typically becoming more complicated in a logical progression.

Synthetic compounds as futuristic antifungal

Due to the increase of resistance to conventional drugs, developing novel anticandidal agents with well-defined mechanisms of action may be a rationalist approach to combating a variety of pathogenic fungal strains. Many pharmacophores and scaffolds have been extensively studied for their antifungal potential. Scaffolds such as the amino acid–derived 1,2-benzisothiazolinone were screened against pathogens that included several dermatophytic fungi [1]. Structure–activity relationships (SARs) revealed the value of a heterocyclic ring, a methyl group, and a phenyl ring for optimal antifungal activity after lead optimization of the scaffold. Despite the fact that these compounds have little structural resemblance to currently available antifungal medications, they show synergy with fluconazole and have fungicidal concentrations comparable to micafungin.

CONTENTS

Introduction . 119

Synthetic compounds as futuristic antifungal..... 119

Olorofim: orotomides (F901318) 120

Natural compounds as promising antifungal..... 120
 Curcumin 121
 Thymol 122
 Berberine 123
 Tetrandrine 124

Antifungals in different stage of clinical trials 125
 Rezafungin 125
 Ibrexafungerp (SCY-078) 125

Fosmanogepix
(APX001).............. 126
Nikkomycin Z 126
Tetrazoles.......... 127
Encochleated
amphotericin B
(MAT2203).......... 127
Aureobasidin A 128
F901318 (olorofim)
129
VL-2397............... 129
T-2307................. 130
AR-12.................. 130

References .. 131

Olorofim: orotomides (F901318)

Olorofim is the first member of the orotomides, a modern antifungal class. The chemical structure of this molecule is depicted. The dihydroorotate dehydrogenase (DHODH), 100 a main enzyme in pyrimidine biosynthesis, is inhibited by this class, which has a novel mechanism of action. Because pyrimidine is involved in the synthesis of DNA, RNA, cell walls, and phospholipids, as well as cell regulation and protein production, 101 it will have a significant impact on fungi. Since the DHODH-target enzyme is present in a number of (fungal) organisms, structural differences may cause different susceptibilities. The IC50 of human and fungus, for example, differs by more than 2000 times [2].

Based on studies that documented the antifungal potentials of such compounds, a new class of azole-based antifungal agents was created by combining two or more biologically important azole scaffolds. To this end, we produced some benzimidazole and 1,3,4-oxadiazole ring hybrid compounds.

Natural compounds as promising antifungal

Fungal infections are at an all-time high, leading to increased morbidity and mortality. The problem is exacerbated by a rise in antimicrobial resistance and populations of patients at risk, as well as a limited range of widely available antifungal medications with numerous side effects. The majorities of clinically used antifungals have toxicity, effectiveness, and cost disadvantages, and their widespread use has resulted in the development of resistant strains. Furthermore, social pressure to cut the use of synthetic fungicides in agriculture has grown over the years. Concerns were raised about the environmental effects as well as the potential health risk related to the use of these antifungal compounds. These disadvantages highlight the need for better and more sophisticated antifungal agents to be developed. Because of their wide range of biological activities, natural products such as crude extracts, essential oils, terpenoids, saponins, phenolic compounds, alkaloids, peptides, and proteins are appealing prototypes for this purpose. The following are a few natural compounds that have been confirmed to be effective against *Candida* species:

Curcumin

Curcumin is the main curcuminoid in turmeric, a common Indian spice that belongs to the ginger family (Zingiberaceae). Desmethoxycurcumin and bis-desmethoxycurcumin are the other two curcuminoids. Turmeric's yellow color is caused by curcuminoids, which are natural phenols. Curcumin comes in a variety of tautomeric variants, including a 1,3-diketo and two enol equivalents. In the solid state and in solution, the enol form is more energetically stable. Curcumin was the most effective at inhibiting/modulating ABCB1, ABCC1, and ABCG2 function. Curcumin can be used as a wide spectrum modulator of MDR, according to recent research. Curcumin is metabolized to dihydrocurcumin (DHC) and tetrahydrocurcumin (THC) by the endogenous reductase mechanism, but it has a poor bioavailability when taken orally [3]. As a result, one of the main metabolites, THC, was tested for its ability to inhibit the three major drug transporters. THC is easily absorbed in the gastrointestinal tract. It was observed that THC also inhibited these transporters, suggesting that metabolite produced from curcumin biotransformation in the body can be used to sensitize MDR cells. Apart from the modulator effects, curcumin has been shown to have potent preclinical antitumor effects [4]. Curcumin's actions on transcription factors, apoptotic genes, angiogenesis regulators, and cellular signaling pathways may be the molecular basis for tumor inhibition [5].

One of the most effective mechanisms for preventing cancer initiation and progression is to inhibit one of these pathways [6]. Curcumin is an ideal compound for developing as a wide spectrum inhibitor because it not only has antitumor properties but also has potent inhibitory effects on ABC drug transporters. Curcumin, a natural polyphenol, has been studied for its antifungal properties against *C. albicans* and non-albicans organisms. Curcumin's inhibitory effects were found to be regardless of the status of multidrug efflux pump proteins belonging to the ABC (ATP-binding cassette transporter) or MFS (major facilitator) transporter superfamilies [7]. Additionally, curcumin is also reported to raise the levels of ROS (reactive oxygen species), which triggered early apoptosis in *Candida* cells. It is known to elevate the expression of stress genes *CAP1* (*C. albicans* AP-1), *CaIPF7817* (putative NADH-dependent flavin oxidoreductase), *SOD2* (superoxide dismutase 2), *GRP2* (NADPH-dependent methyl glyoxal reductase), and *CAT1* (catalase 1) [8].

Thymol

Thymol [5-methyl-2-(1-methylethyl) phenol] is the most important component of thyme oil, and it is structurally similar to carvacrol but with a different hydroxyl group on the phenolic ring [9]. It is clear that thymol interacts with membrane permeability, resulting in membrane potential loss, K+ ion leakage, and ATP and carboxy fluorescein leakage [10]. Furthermore, thymol interacts with membrane-bound or periplasmic proteins through hydrophilic and hydrophobic interactions. It's a naturally occurring phenol that's been used to treat a variety of fungal diseases. Thymol has a number of medicinal qualities, including antibacterial and antifungal activities, and its activity against biofilms has been studied and reported extensively [11].

Berberine

Berberine is an isoquinoline alkaloid that is a quaternary ammonium salt. Berberine is bright yellow color, which is why berberis species were used to dye fur, leather, and wood in the past. Berberine is still used to dye wool in Northern India today [12]. Berberine, on the other hand, is widely regarded as an ineffective antibiotic, though some argue that when combined with other biochemical substances derived from the same plants, it may be useful [13].

Inhibition of endoplasmic reticulum stress appears to be a key mechanism by which berberine inhibits the HIV PI-induced inflammatory response in macrophages, according to studies. According to some sources, berberine can be used as a complementary therapy for HIV infection. There are findings suggesting that the combination of fluconazole and berberine produced potently synergistic action against fluconazole-resistant *Candida albicans* in vitro [14]. This interaction may be due to berberine inhibition of sterol 24-methyl transferase. Furthermore, active efflux of drugs (with overexpressed CDR1, MDR1, or FLU1) and target enzyme alterations have been identified as mechanisms for azole resistance. Berberine can have a synergistic effect with fluconazole by influencing a resistance factor [14]. Berberine is a promising and safe agent against fluconazole-resistant *C. albicans*, in vitro when used in combination with fluconazole, but further research is required to establish the underlying mechanism of the synergistic action.

Tetrandrine

Tetrandrine, a calcium channel blocker, is a bis-benzylisoquinoline alkaloid. It has antiinflammatory, antiimmune, and antiallergenic properties. It prevents mast cells from degranulating. It has an antiarrhythmic effect that is similar to quinidine. It was discovered in the Menispermaceae family's Stephania tetrandra S Moore, as well as other Chinese and Japanese herbs. It has vasodilatory properties, which means it can lower blood pressure. Tetrandrine has the ability to be used to treat liver disease and cancer. Tetrandrine can be useful in preventing conjunctival scarring and fibrosis after trabeculectomy or in patients with serious conjunctival inflammation.

Tetrandrine has antiinflammatory and antifibrogenic actions, which make tetrandrine and related compounds potentially useful in the treatment of lung silicosis, liver cirrhosis, and rheumatoid arthritis. There are reports, which demonstrate that tetrandrine can increase the sensitivity of *C. albicans* of fluconazole in vitro at noncytotoxic doses. Tetrandrine and associated compounds have antiinflammatory and antifibrogenic properties, making them particularly useful in the treatment of lung silicosis, liver cirrhosis, and rheumatoid arthritis. There have been reports that tetrandrine can increase fluconazole sensitivity in *C. albicans* in vitro at noncytotoxic doses.

Antifungals in different stage of clinical trials
Rezafungin

Rezafungin (formerly CD101) is a novel echinocandin in clinical development that has exceptional stability and solubility, as well as a long half-life that allows for front-loaded drug exposure with once-weekly dosing. Rezafungin has been shown to have efficacy against *Candida* spp. and *Aspergillus* spp., including subsets of echinocandin-resistant *Candida auris* and azole-resistant *Aspergillus* isolates, comparable to other echinocandins [15]. It is currently in the phase 3 postproduction for the treatment of candidemia and invasive candidiasis. For its development program, the US Food and Drug Administration (FDA) has designated intravenous (IV) rezafungin as a Qualified Infectious Disease Product (QIDP) with fast track status [16].

30

Ibrexafungerp (SCY-078)

Ibrexafungerp (previously MK-3118 and SCY-078; developed by Scynexis, Jersey City, NJ) is a semisynthetic derivative of enfumafungin that is a first-in-class oral glucan synthase inhibitor. Ibrexafungerp works against *Candida* spp. by inhibiting (13)-D-glucan synthase, a central component of the fungal cell wall. Ibrexafungerp is being formulated for both oral and IV administration, but only the oral administration is undergoing clinical trials at this time [17].

Fosmanogepix (APX001)

APX001A is a novel antifungal that targets the conserved Gwt1 enzyme required for localization of glycosylphosphatidylinositol (GPI) thus inactivates posttranslational modification of their anchor proteins, also known as mannoproteins [18]. The GPI-anchored proteins bind to −1,3-glucan and help preserve the fungal cell wall's integrity, as well as play a role in adherence and invasion of host tissues [19]. When GPI-anchor synthesis is disrupted, −1,3-glucan is exposed, increasing immune cell recognition of the fungus. Several phase 3 trials are currently underway to confirm the efficacy and toxicity pertaining to its role against pathogenic fungi [19].

Nikkomycin Z

Nikkomycin Z (NikZ) has fungicidal efficacy against certain fungal pathogens, but patients must undergo long-term treatment, often for years [20]. Its first-in-class antifungal is derived from

Streptomyces tendae that works by inhibiting chitin synthases, an important component of the fungal cell wall. NikZ is selective to fungal cells and has little to no toxicity in humans since the target enzyme is absent in mammalian hosts [21].

Tetrazoles

Tetrazoles are next-generation azoles that target the clinically important drug-drug interactions caused by inhibition of off-target CYP450; this is one of the major drawbacks of the azole class of antifungals. VT-1129 [22], VT-1161 (or oteseconazole), and VT-1598 [23] (developed by Viamet Pharmaceuticals, Durham, NC) are attempting to overcome this limitation by improving binding discrimination between fungal and mammalian CYP450 enzymes [24].

Encochleated amphotericin B (MAT2203)

Amphotericin B (AmB) is a broad-spectrum fungicide that kills a wide variety of yeasts and molds. The biggest disadvantage of the current formulations is their toxicity, as well as the higher costs associated with better tolerated lipid formulations. The oral amphotericin B (CAmB) formulation MAT2203 (developed by MatinasBioPharma, Bedminster, NJ) is a novel encochleated amphotericin B (CAmB) formulation. Encochleated AmB, like all types of amphotericin, binds to sterols, increasing ion permeability across the cell membrane [25]. In the interior of a calcium-phospholipid anhydrous crystal, the novel nanoparticle-based encochleated AmB (CAmB) formulation encapsulates, preserves, and delivers the cargo molecule AmB.

Aureobasidin A

Aureobasidin A (AbA) is an antifungal that targets the essential inositol phosphoryl ceramide that targets a broad spectrum of pathogens like *Saccharomyces cerevisiae*, *Candida* spp., *Cryptococcus* spp., *Schizosaccharomyces pombe*, and certain *Aspergillus* spp. [26]. While this drug is still in the preclinical stage of development, it has a promising clinical profile [27].

F901318 (olorofim)

F901318 (olorofim) is a new antifungal drug with a high antifungal activity against *Aspergillus* organisms [28]. F901318 is an antifungal drug in clinical development that shows excellent efficacy against a wide variety of dimorphic and filamentous fungi. It is the leading representative of a novel class of drugs known as orotomides. In the de novo pyrimidine biosynthesis pathway, it targets dihydroorotate dehydrogenase (DHODH) [29].

VL-2397

VL-2397 (formerly ASP2397, developed by Vical Pharmaceuticals) may be the start of a new antifungal drug class. It's a cyclic hexapeptide extracted from *Acremonium persicinum* fermentation broth. The siderophore ferrichrome, a low molecular weight siderophore with high specificity for iron, is structurally similar to VL-2397. Its impact on an unknown intracellular target causes activity [30].

T-2307

T-2307 is an investigational arylamidine that is structurally similar to a class of aromatic diamidines that includes pentamidine (and was synthesized at Toyama Chemical Co., Llt., Tokyo, Japan) [31] The process by which T-2307 inhibits fungus mitochondrial function is incompletely understood, and unraveling it could help researchers develop new antifungal drugs. T-2307 has the potential to be a powerful injectable treatment for candidiasis, cryptococcosis, and aspergillosis [32].

AR-12

AR-12 (Flemington, NJ-based Arno Therapeutics Inc.) also called OSU-03012 is an antitumor celecoxib derivative that has advanced to Phase I clinical trials as an anticancer agent and has activity against fungi, bacteria, and viruses [33]. In vitro, AR-12 inhibits fungal acetyl coenzyme A (acetyl-CoA) synthetase and is fungicidal at concentrations comparable to those found in human plasma [34].

References

[1] B. Hube, R. Hay, J. Brasch, S. Veraldi, M. Schaller, Dermatomycoses and inflammation: the adaptive balance between growth, damage, and survival, Journal de Mycologie Médicale 25 (2015) e44–e58, https://doi.org/10.1016/j.mycmed.2014.11.002.

[2] G.H. Kathwate, R.B. Shinde, S. Mohan Karuppayil, Non-antifungal drugs inhibit growth, morphogenesis and biofilm formation in Candida albicans, The Journal of Antibiotics 74 (2021) 346–353, https://doi.org/10.1038/s41429-020-00403-0.

[3] J. Murugesh, R.G. Annigeri, G.K. Mangala, P.H. Mythily, J. Chandrakala, Evaluation of the antifungal efficacy of different concentrations of Curcuma longa on Candida albicans: an in vitro study, Journal of Oral and Maxillofacial Pathology 23 (2019) 305, https://doi.org/10.4103/jomfp.JOMFP_200_18.

[4] E. Chen, B. Benso, D. Seleem, L.E.N. Ferreira, S. Pasetto, V. Pardi, R.M. Murata, Fungal-host interaction: curcumin modulates proteolytic enzyme activity of Candida albicans and inflammatory host response in vitro, International Journal of Dentistry 2018 (2018) 2393146, https://doi.org/10.1155/2018/2393146.

[5] D.B. Singh, M.K. Gupta, D.V. Singh, S.K. Singh, K. Misra, Docking and in silico ADMET studies of noraristeromycin, curcumin and its derivatives with Plasmodium falciparum SAH hydrolase: a molecular drug target against malaria, Interdisciplinary Sciences: Computational Life Sciences 5 (2013) 1–12, https://doi.org/10.1007/s12539-013-0147-z.

[6] B.B. Aggarwal, A. Kumar, A.C. Bharti, Anticancer potential of curcumin: preclinical and clinical studies, Anticancer Research 23 (2003) 363–398.

[7] M. Sharma, R. Manoharlal, S. Shukla, N. Puri, T. Prasad, S.V. Ambudkar, R. Prasad, Curcumin modulates efflux mediated by yeast ABC multidrug transporters and is synergistic with antifungals, Antimicrobial Agents and Chemotherapy 53 (2009) 3256–3265, https://doi.org/10.1128/AAC.01497-08.

[8] A. Kumar, S. Dhamgaye, I.K. Maurya, A. Singh, M. Sharma, R. Prasad, Curcumin targets cell wall integrity via calcineurin-mediated signaling in candida albicans, Antimicrobial Agents and Chemotherapy 58 (2014) 167–175, https://doi.org/10.1128/AAC.01385-13.

[9] N. Guo, J. Liu, X. Wu, X. Bi, R. Meng, X. Wang, H. Xiang, X. Deng, L. Yu, Antifungal activity of thymol against clinical isolates of fluconazole-sensitive and -resistant Candida albicans, Journal of Medical Microbiology 58 (2009) 1074–1079, https://doi.org/10.1099/jmm.0.008052-0.

[10] R.D. de Castro, T.M.P.A. de Souza, L.M.D. Bezerra, G.L.S. Ferreira, E.M.M. de Brito Costa, A.L. Cavalcanti, Antifungal activity and mode of action of thymol and its synergism with nystatin against Candida species involved with infections in the oral cavity: an in vitro study, BMC Complementary and Alternative Medicine 15 (2015) 417, https://doi.org/10.1186/s12906-015-0947-2.

[11] H. Jafri, I. Ahmad, Thymus vulgaris essential oil and thymol inhibit biofilms and interact synergistically with antifungal drugs against drug resistant strains of Candida albicans and Candida tropicalis, Journal de Mycologie Médicale. 30 (2020) 100911, https://doi.org/10.1016/j.mycmed.2019.100911.

[12] S. Dhamgaye, F. Devaux, P. Vandeputte, N.K. Khandelwal, D. Sanglard, G. Mukhopadhyay, R. Prasad, Molecular mechanisms of action of herbal antifungal alkaloid berberine, in Candida albicans, PLoS One 9 (2014) e104554, https://doi.org/10.1371/journal.pone.0104554.

[13] A.R. da Silva, J.B. de Andrade Neto, C.R. da Silva, R.D.S. Campos, R.A. Costa Silva, D.D. Freitas, F.B.S.A. do Nascimento, L.N.D. de Andrade, L.S. Sampaio, T.B. Grangeiro, H.I.F. Magalhães, B.C. Cavalcanti, M.O. de Moraes, H.V. Nobre Júnior, Berberine antifungal activity in fluconazole-resistant pathogenic yeasts: action mechanism evaluated by flow cytometry and biofilm growth inhibition in Candida spp, Antimicrob Agents Chemother 60 (2016) 3551–3557, https://doi.org/10.1128/AAC.01846-15.

[14] H. Quan, Y.-Y. Cao, Z. Xu, J.-X. Zhao, P.-H. Gao, X.-F. Qin, Y.-Y. Jiang, Potent in vitro synergism of fluconazole and berberine chloride against clinical isolates of Candida albicans resistant to fluconazole, Antimicrobial Agents and Chemotherapy 50 (2006) 1096–1099, https://doi.org/10.1128/AAC.50.3.1096-1099.2006.

[15] Y.Y. Ham, J.S. Lewis, G.R. Thompson, Rezafungin: a novel antifungal for the treatment of invasive candidiasis, Future Microbiol 16 (2021) 27–36, https://doi.org/10.2217/fmb-2020-0217.

[16] M. Helleberg, K.M. Jørgensen, R.K. Hare, R. Datcu, A. Chowdhary, M.C. Arendrup, Rezafungin *in vitro* activity against contemporary nordic clinical *Candida* isolates and *Candida auris* determined by the EUCAST reference method, Antimicrobial Agents and Chemotherapy 64 (2020) e02438−19, https://doi.org/10.1128/AAC.02438-19.

[17] M.R. Davis, M.A. Donnelley, G.R. Thompson, Ibrexafungerp: a novel oral glucan synthase inhibitor, Medical Mycology 58 (2020) 579−592, https://doi.org/10.1093/mmy/myz083.

[18] S. Alkhazraji, T. Gebremariam, A. Alqarihi, Y. Gu, Z. Mamouei, S. Singh, N.P. Wiederhold, K.J. Shaw, A.S. Ibrahim, Fosmanogepix (APX001) is effective in the treatment of immunocompromised mice infected with invasive pulmonary scedosporiosis or disseminated fusariosis, Antimicrobial Agents and Chemotherapy 64 (2020), https://doi.org/10.1128/AAC.01735-19.

[19] N.P. Wiederhold, L.K. Najvar, K.J. Shaw, R. Jaramillo, H. Patterson, M. Olivo, G. Catano, T.F. Patterson, Efficacy of delayed therapy with fosmanogepix (APX001) in a murine model of Candida auris invasive candidiasis, Antimicrobial Agents and Chemotherapy 63 (2019), https://doi.org/10.1128/AAC.01120-19.

[20] D.J. Larwood, Nikkomycin Z—ready to meet the promise? Journal of Fungi 6 (2020) 261, https://doi.org/10.3390/jof6040261.

[21] D.E. Nix, R.R. Swezey, R. Hector, J.N. Galgiani, Pharmacokinetics of nikkomycin Z after single rising oral doses, Antimicrobial Agents and Chemotherapy 53 (2009) 2517−2521, https://doi.org/10.1128/AAC.01609-08.

[22] E. Łukowska-Chojnacka, A. Kowalkowska, M. Gizińska, M. Koronkiewicz, M. Staniszewska, Synthesis of tetrazole derivatives bearing pyrrolidine scaffold and evaluation of their antifungal activity against Candida albicans, European Journal of Medicinal Chemistry 164 (2019) 106−120, https://doi.org/10.1016/j.ejmech.2018.12.044.

[23] M. Bondaryk, E. Łukowska-Chojnacka, M. Staniszewska, Tetrazole activity against Candida albicans. The role of KEX2 mutations in the sensitivity to (±)-1-[5-(2-chlorophenyl)-2H-tetrazol-2-yl]propan-2-yl acetate, Bioorganic & Medicinal Chemistry Letters 25 (2015) 2657−2663, https://doi.org/10.1016/j.bmcl.2015.04.078.

[24] M. Staniszewska, M. Gizińska, E. Mikulak, K. Adamus, M. Koronkiewicz, E. Łukowska-Chojnacka, New 1,5 and 2,5-disubstituted tetrazoles-dependent activity towards surface barrier of Candida albicans, European Journal of Medicinal Chemistry 145 (2018) 124−139, https://doi.org/10.1016/j.ejmech.2017.11.089.

[25] M. Aigner, C. Lass-Flörl, Encochleated amphotericin B: is the oral availability of amphotericin B finally reached? Journal of Fungi (Basel) 6 (2020) https://doi.org/10.3390/jof6020066.

[26] W. Zhong, M.W. Jeffries, N.H. Georgopapadakou, Inhibition of inositol phosphorylceramide synthase by aureobasidin A in Candida and Aspergillus species, Antimicrobial Agents and Chemotherapy 44 (2000) 651−653, https://doi.org/10.1128/aac.44.3.651-653.2000.

[27] A.Q.I. Alqaisi, A.J. Mbekeani, M.B. Llorens, A.P. Elhammer, P.W. Denny, The antifungal Aureobasidin A and an analogue are active against the protozoan parasite Toxoplasma gondii but do not inhibit sphingolipid biosynthesis, Parasitology 145 (2018) 148−155, https://doi.org/10.1017/S0031182017000506.

[28] J.D. Oliver, G.E.M. Sibley, N. Beckmann, K.S. Dobb, M.J. Slater, L. McEntee, S. du Pré, J. Livermore, M.J. Bromley, N.P. Wiederhold, W.W. Hope, A.J. Kennedy, D. Law, M. Birch, F901318 represents a novel class of antifungal drug that inhibits dihydroorotate dehydrogenase, Proceedings of the National Academy of Sciences 113 (2016) 12809−12814, https://doi.org/10.1073/pnas.1608304113.

[29] S. du Pré, N. Beckmann, M.C. Almeida, G.E.M. Sibley, D. Law, A.C. Brand, M. Birch, N.D. Read, J.D. Oliver, Effect of the novel antifungal drug F901318 (olorofim) on growth and viability of Aspergillus fumigatus, Antimicrobial Agents and Chemotherapy 62 (2018), https://doi.org/10.1128/AAC.00231-18.

[30] A.-M. Dietl, M. Misslinger, M.M. Aguiar, V. Ivashov, D. Teis, J. Pfister, C. Decristoforo, M. Hermann, S.M. Sullivan, L.R. Smith, H. Haas, The siderophore transporter Sit1 determines susceptibility to the antifungal VL-2397, Antimicrobial Agents and Chemotherapy 63 (2019), https://doi.org/10.1128/AAC.00807-19.

[31] K. Yamashita, T. Miyazaki, Y. Fukuda, J. Mitsuyama, T. Saijo, S. Shimamura, K. Yamamoto, Y. Imamura, K. Izumikawa, K. Yanagihara, S. Kohno, H. Mukae, The novel arylamidine T-2307 selectively disrupts yeast mitochondrial function by inhibiting respiratory chain complexes, Antimicrobial Agents and Chemotherapy 63 (2019), https://doi.org/10.1128/AAC.00374-19.

[32] J. Mitsuyama, N. Nomura, K. Hashimoto, E. Yamada, H. Nishikawa, M. Kaeriyama, A. Kimura, Y. Todo, H. Narita, In vitro and in vivo antifungal activities of T-2307, a novel arylamidine, Antimicrobial Agents and Chemotherapy 52 (2008) 1318−1324, https://doi.org/10.1128/AAC.01159-07.

[33] K. Koselny, J. Green, L. Favazzo, V.E. Glazier, L. DiDone, S. Ransford, D.J. Krysan, Antitumor/antifungal celecoxib derivative AR-12 is a nonnucleoside inhibitor of the ANL-family adenylating enzyme acetyl CoA

synthetase, ACS Infectious Diseases 2 (2016) 268–280, https://doi.org/10.1021/acsinfecdis.5b00134.

[34] K. Koselny, J. Green, L. DiDone, J.P. Halterman, A.W. Fothergill, N.P. Wiederhold, T.F. Patterson, M.T. Cushion, C. Rappelye, M. Wellington, D.J. Krysan, The celecoxib derivative AR-12 has broad-spectrum antifungal activity in vitro and improves the activity of fluconazole in a murine model of cryptococcosis, Antimicrobial Agents and Chemotherapy 60 (2016) 7115–7127, https://doi.org/10.1128/AAC.01061-16.

CHAPTER 9

Drug repurposing for development of effective anticandidals

Introduction

The concept of testing medications that have already been approved for a specific use to see if they can be used for a different purpose is not new to science. This concept has recently gained traction as a result of recent advances in computational sciences and the ability to process massive data sets on a high-performance computing unit. In the case of invasive fungal infections, drug repurposing may provide successful treatments. Target-based repurposing, in which a drug interacts with a gene or protein, and disease-based repurposing, in which a drug is linked to new indications, are the two approaches offered by computational repurposing [1]. Newer drug research and development has a long track record before it reaches the market. The drug development involves four major aspects of evaluation.

CONTENTS

Introduction . 137
References .. 144

1. Evaluation of toxicity of drug,
2. Target validation
3. Hit-to-lead optimization
4. Metabolism and pharmacokinetic pathways in vivo.

It takes about one to two decades for a drug to pass through the 11 landmarks to reach the market whereas a repurposed drug might cover these mile stones in a span of 3–12 years. Although the time to the clinics is reduced manifold but drug repurposing comes with its drawbacks. For example, the drug that is being repurposed needs to be tested for its efficacy in a new dosage that may be very high when tested to be used as antimicrobial compared to its current dosage. Also, this new dosage might affect the pharmacodynamics and pharmacokinetics for the drug. One major challenge

Anticandidal Therapeutics. https://doi.org/10.1016/B978-0-443-18744-5.00008-3
Copyright © 2023 Elsevier Inc. All rights reserved.

in repurposing a drug is its dosage which should be similar to that for established usage. Compatibility between human pharmacokinetic profile and pharmacokinetics and pharmacodynamics of the drug is another key factor for the application of a drug other than its intended use. A good way is that a combination of drugs can be used to circumvent the harmful effects of increased dosage of repurposed drugs. But this also means that the interaction of new drug combination should be as effective and tested for its clinical application. This might nullify the time advantage acquired with repurposed drugs. Also, multiple routes of application, i.e., systemic, topical, or inhalation can be looked for as a good option. Along with these clinical challenges lie the more legal issues of patents and marketing.

Nevertheless, drug repurposing is a powerful weapon for researchers and scientists as an accelerated method for drug development. Recently, the easy availability of data from concerned libraries and the never ending development of new techniques for screening have proven to be a boon for this. When combined with screening techniques, such breakthroughs could pave the way for the long-awaited discovery of known and approved antimicrobial drugs. Below, we'll go through some of the methods that have been proposed to speed up the selection of the most promising repurposing candidates. Traditional empirical screening, experimental screening assays, and d screening are only a few examples.

Screening based on evidence—The systematic screening of nonantimicrobial FDA-approved drugs for antimicrobial activity in cell-based models is a popular approach for drug repurposing. These empirical phenotypic-based screens ask a simple question: The molecules inhibit the growth of the microorganism of interest? Traditional repurposing techniques, on the other hand, often involve a prior awareness of disease or binding characteristics, as well as the mechanism of action (MOA). However, phenotype-based screening offers little or no insight into the MOA, which is a major impediment to modern drug development efforts, necessitating further research into mechanistic analysis to establish exactly how the drug is functioning to achieve the desired impact [2]. The known functions of the identified drugs may provide hints for the analysis of the MOA in some cases. The defined properties of the known drugs may provide hints for studying the MOA in certain cases. The design of optimized next-generation formulations can also benefit from in-depth analysis of the MOA and molecular targets. MOA studies, however, can be important, particularly in the case of antimicrobial discovery, since they help researchers to understand the molecular and metabolic basis of growth inhibition, as well as how microorganisms can respond to such alteration. It is important to have this concept in process of designing effective resistance-reduction strategies.

Researchers are increasingly shifting to unconventional screening methods which go beyond in vitro models that test pathogen growth inhibition. This includes the following: target-based phenotypic screens and screens in more clinically specific settings are examples of platforms. Target-based phenotypic screens, whole-animal screens, and screening under conditions that better represent the environment pathogens encounter during infection have all been used for antimicrobial drug repurposing. Target-based phenotypic screening techniques have used sophisticated monitors to detect direct or indirect inhibitors of a mechanism of interest, pathway of interest, or image-based assays. Increasingly, in vivo screening assays with relatively high screening throughput are being produced which are best used with whole organism models.

Antibiotics combined with other drugs is an appealing choice for novel drug production, as it extends the antimicrobial drug's lifespan and overcomes the problem of resistance to antibacterial drugs. Among the three key methods that have been used first is the quest for synergizing compounds, in which repurposed drugs have associated MOAs that improve the antimicrobial agent's action. Synergistic combinations are seen to be extremely effective and medicinally specific. But synergy can have two opposing effects on resistance: it can slow the development of resistance by clearing the infection earlier, reducing the amount of time for resistant mutations to emerge, but it can also boost the selective advantage of single-drug resistant mutants [3]. The second strategy is the quest for drugs that target resistance mechanisms (such as reversing efflux and inhibiting resistance enzymes) to resensitize resistant microorganisms [4]. Because of the clinical success of antibiotic–adjuvant combinations like augmentin, the adjuvant approach is a very appealing avenue for the identification of innovative therapeutics [5]. Notably, restoring or potentiating the efficacy of antimicrobial drugs provides an additional therapeutic choice at a time when treatment options are scarce. The last but effective means of drug repurposing is the combinatorial strategy [6]. It is the quest for drugs that allows for increased action of the other, in which both agents are ineffective on their own but one reveals the target for the other. Nevertheless, one of the most significant challenges in studying the effects of drug combinations is the large number of studies needed to systematically investigate all potential drug combinations.

Prescreening or in silico profiling of compound libraries may also be a good way to find potential drug leads for infectious diseases [7]. These approaches rely on high performance computing and a large array of publicly accessible pharmacological, biological, and chemical data. In most cases, in silico screening is dependent on ligands or networks. Molecular docking, which predicts a molecule's orientation inside a protein target's binding site based on the assumption that similar binding sites bind similar molecules is a popular computational platform for drug discovery [8]. Based on the high degree of binding site similarity between human catechol-O-methyltransferase and the bacterial enoyl-acyl carrier protein reductase (InhA), an enzyme necessary for fatty acid synthesis, entacapone, a medication used to treat Parkinson disease, was discovered as a potential antibacterial compound against multiple drug-resistant *Mycobacterium tuberculosis*. Using libraries of previously approved drugs, other ligand-based pharmacophore-modeling studies have found virulence inhibitors of Methicillin-resistant *Staphylococcus aureus*

(MRSA), efflux pump inhibitors in *S. aureus*, and a drug-resistant Gram-negative pathogen inhibitor that affects galactose metabolism and lipopolysaccharide biosynthesis [9]. The methods of systems engineering are used in network-based in silico approaches [10]. The use of network-based computational approaches to repurpose drugs for infectious diseases is becoming more appealing [11]. In this paradigm, the approaches used differ. It applies the biology and bioinformatics as tools to compare host responses to drugs and diseases. Lists of over- and underexpressed genes in a biological system add complexity from simple interaction networks to more complex interaction networks. This type of information can then be compared from a variety of high-throughput techniques to put in place that the existing medications and new disease indications have therapeutic relationships.

The computational drug repurposing in fungi is yet to see its first success story with a compound hitting the shelves; experimental evidence is mounting in favor of the approach's viability. The capacity for cost-effective processing of compound and data sets, orders of magnitude greater than what can be done in the laboratory, is the allure of virtual screening. On the other hand, this method is highly reliant on the availability of high-quality data for prediction. As a result, it is possible that it will skip a lot of high-potential compounds. When screening small libraries, false negative findings are more difficult to detect and can hinder attempts to identify drug interactions. More replication experiments and robust statistical analysis as well as a variety of biological assays (for example, testing in more cell lines) will limit the number of false negative results; however, these choices would be affected by time and cost limitations [9]. None of these approaches would likely be sufficient to reveal or model the complex interactions that exist between drugs, targets, and diseases. As a result, it is thought that greater integration and use of different computational approaches would greatly help repurposing efforts.

High-throughput screenings/bioassays are still being used to discover novel drugs and/or establish molecular targets of newly discovered antifungal agents [12]. This is particularly true when it comes to determining the role of specific genes, genetic pathways, or previously undetected lipid changes in cellular membranes, cross-talks between lipid molecules and mitochondrial dysfunction [13], cell wall integrity and filamentous fungal development, and other factors that could explain resistance to antifungal drugs [14]. Candidiasis is one example of fungal infections showing increased incidences of fungal resistance to a class of azoles, and this has become a global human health problem. It is particularly dangerous for immunocompromised people and patients with lung diseases. Not just *Candida*-based infections but other infections like aspergillosis are also severely persistent. Among the various forms of aspergillosis that have been reported (including allergic bronchopulmonary aspergillosis, allergic *Aspergillus* sinusitis, aspergilloma, chronic pulmonary aspergillosis, invasive aspergillosis (IA), and cutaneous aspergillosis), allergic bronchopulmonary aspergillosis is one of the most common. As a consequence, there is a constant need to boost the efficacy of currently available antifungal drugs or to establish new intervention strategies in not just anticandidal infection but overall antifungal therapeutics.

While drug repurposing has become a viable strategy for accelerating the production of new antifungal drugs, it still necessitates the use of highly sensitive screening systems. Antifungal "chemosensitization" has been introduced as a modern intervention technique in which the coapplication of a second compound (namely, a chemosensitizer; natural or synthetic) with a commercial medication has been found to improve the antifungal efficacy of the coapplied drug [15]. Chemosensitization has the distinct advantage that unlike combination therapy (i.e., the simultaneous administration of two or more commercial antifungal drugs) a chemosensitizer does not require a high degree of antifungal potency [16]. Instead, by modulating the pathogen's protection mechanism, a chemosensitizer allows the target pathogen to become more vulnerable to the commercial coapplied medication. The chemosensitization method and/or the inclusion of fungal mutants missing key genes in the cellular targets could improve the sensitivity of the drug repurposing process.

In a recent study repurposing FDA approved drugs against the human fungal pathogen by Kim et al., *C. albicans* was killed by 15 medications that were initially licensed to treat a variety of infectious and noninfectious diseases [9]. Additionally, one of those drugs, Octodrine, has antimicrobial activity over a broad spectrum. Octodrine was found to be a highly effective anti-*Candida* drug in destroying serum-grown *C. albicans* without disturbing normal functioning of human MP (macrophages) and skin cells as compared to other anti-*Candida* drugs.

Antifungal activity of mycophenolic acid, disulfiram, fluvastatin, and octodrine has been found in medicines used for a variety of purposes including immune suppression, antihyperlipidemics, and decongestant [9]. These are the most important class of drugs from the standpoint of drug repurposing, as they demonstrate efficacy against a number of diseases but no identified antifungal activity till this date. There are three chemical groups of these drugs: azole (fluvastatin) [17], oxole (mycophenolic acid) [18], and other structures (disulfiram and octodrine) [19]. These four medications, which were previously licensed to treat noninfectious diseases, but now have antifungal properties, may be repurposed as new antifungals.

In addition, a few antimicrobial/antiseptic drugs have been shown in preventing *C. albicans* growth. Such medications contain antifungal formulations, and they are made up of three chemical classes: oxoles (nifuroxime), pyridines (nitroxoline and chlorquinaldole), and other structures (octanoic acid and benzethonium chloride). Azoles (fluconazole, captan, clotrimazole, and miconazole), pyridine (pyrithione zinc), and other structures are antifungal medicines (antimycin A) have shown to kill *C. albicans* in the screening approach. The fact that these other FDA-approved drugs for treating nonfungal infectious diseases were also found to destroy *C. albicans* suggests that they may be repurposed as broad-spectrum antimicrobial.

References

[1] A. Jha, A. Kumar, Anticandidal agent for multiple targets: the next paradigm in the discovery of proficient therapeutics/overcoming drug resistance, Future Medicinal Chemistry 11 (2019) 2955–2974, https://doi.org/10.4155/fmc-2018-0479.

[2] I.D. Jacobsen, Fungal infection strategies, Virulence 10 (2019) 835–838, https://doi.org/10.1080/21505594.2019.1682248.

[3] S. Liu, L. Yue, W. Gu, X. Li, L. Zhang, S. Sun, C. Pierce, J. Lopez-Ribot, C. Orasch, O. Marchetti, J. Garbino, J. Schrenzel, S. Zimmerli, K. Muhlethaler, F. Guo, Y. Yang, Y. Kang, B. Zang, W. Cui, B. Qin, M. Pfaller, P. Rhomberg, S. Messer, R. Jones, M. Castanheira, J. Eddouzi, J. Parker, L. Vale-Silva, A. Coste, F. Ischer, S. Kelly, T. Chen, Y. Chen, Y. Chen, P. Lu, M. Sharma, D. Biswas, A. Kotwal, B. Thakuria, B. Kakati, B. Chauhan, S. Liu, Y. Hou, X. Chen, Y. Gao, H. Li, S. Sun, M. Azevedo, R. Teixeira-Santos, A. Silva, L. Cruz, E. Ricardo, C. Pina-Vaz, Q. Yu, C. Xiao, K. Zhang, C. Jia, X. Ding, B. Zhang, Q. Yu, X. Ding, B. Zhang, N. Xu, C. Jia, J. Mao, A. Rodrigues, C. Pina-Vaz, P. Mardh, J. Martinez-de-Oliveira, A. Freitas-da-Fonseca, E. Krajewska-Kulak, W. Niczyporuk, J. Afeltra, R. Vitale, J. Mouton, P. Verweij, Q. Yu, X. Ding, N. Xu, X. Cheng, K. Qian, B. Zhang, N. Bulatova, R. Darwish, S. Liu, Y. Hou, W. Liu, C. Lu, W. Wang, S. Sun, F. Liu, L. Pu, Q. Zheng, Y. Zhang, R. Gao, X. Xu, Y. Chen, S. Yu, H. Huang, Y. Chang, V. Lehman, F. Silao, J. Zhang, F. Silao, U. Bigol, A. Bungay, M. Nicolas, J. Heitman, J. Blankenship, J. Heitman, Y. Chen, A. Brand, E. Morrison, F. Silao, U. Bigol, M. France Jr, C. Patenaude, Y. Zhang, B. Cormack, J. Kohler, R. Rao, Q. Yu, F. Wang, Q. Zhao, J. Chen, B. Zhang, X. Ding, C. da Silva, N.J. de Andrade, J. Sidrim, M. Angelo, H. Magalhaes, B. Cavalcanti, T. Bagar, M. Bencina, J. Teng, R. Goto, K. Iida, I. Kojima, H. Iida, K. Golabek, J. Strzelczyk, A. Owczarek, P. Cuber, A. Slemp-Migiel, A. Wiczkowski, S. Lockhart, C. Bolden, N. Iqbal, R. Kuykendall, S. Sun, Y. Li, Q. Guo, C. Shi, J. Yu, L. Ma, M. Pfaller, D. Diekema, F. Odds, W. Shi, Z. Chen, X. Chen, L. Cao, P. Liu, S. Sun, Q. Guo, S. Sun, J. Yu, Y. Li, L. Cao, Y. Gao, C. Zhang, C. Lu, P. Liu, Y. Li, H. Li, L. Sun, K. Liao, S. Liang, P. Yu, D. Wang, M. Pfaller, D. Sheehan, J. Rex, J. Cui, J. Kaandorp, P. Sloot, C. Lloyd, M. Filatov, S. Yu, Y. Chang, Y. Chen, Q. Yu, B. Zhang, B. Yang, J. Chen, H. Wang, C. Jia, R. Jensen, K. Astvad, L. Silva, D. Sanglard, R. Jorgensen, K. Nielsen, Synergistic effect of fluconazole and calcium channel blockers against resistant Candida albicans, PLoS One 11 (2016) e0150859, https://doi.org/10.1371/journal.pone.0150859.

[4] G. Szakács, J.K. Paterson, J.A. Ludwig, C. Booth-Genthe, M.M. Gottesman, Targeting multidrug resistance in cancer, Nature Reviews Drug Discovery 5 (2006) 219–234, https://doi.org/10.1038/nrd1984.

[5] P. Bernal, C. Molina-Santiago, A. Daddaoua, M.A. Llamas, Antibiotic adjuvants: identification and clinical use, Microbial Biotechnology 6 (2013) 445–449, https://doi.org/10.1111/1751-7915.12044.

[6] M. Spitzer, N. Robbins, G.D. Wright, Combinatorial strategies for combating invasive fungal infections, Virulence 8 (2017) 169–185, https://doi.org/10.1080/21505594.2016.1196300.

[7] C. Harrison, Phenotypic screening: a more rapid route to target deconvolution, Nature Reviews Drug Discovery 13 (2014) 102–103, https://doi.org/10.1038/nrd4243.

[8] T. Kaur, A. Madgulkar, M. Bhalekar, K. Asgaonkar, Molecular docking in formulation and development, Current Drug Discovery Technologies 16 (2019) 30–39, https://doi.org/10.2174/1570163815666180219112421.

[9] K. Kim, L. Zilbermintz, M. Martchenko, Repurposing FDA approved drugs against the human fungal pathogen, Candida albicans, Annals of Clinical Microbiology and Antimicrobials 14 (2015) 32, https://doi.org/10.1186/s12941-015-0090-4.

[10] S. Alaimo, A. Pulvirenti, Network-based drug repositioning: approaches, resources, and research directions, Methods in Molecular Biology 1903 (2019) 97–113, https://doi.org/10.1007/978-1-4939-8955-3_6.

[11] Y. Zhou, Y. Hou, J. Shen, Y. Huang, W. Martin, F. Cheng, Network-based drug repurposing for novel coronavirus 2019-nCoV/SARS-CoV-2, Cell Discovery 6 (2020) 14, https://doi.org/10.1038/s41421-020-0153-3.

[12] M.A. Farha, E.D. Brown, Drug repurposing for antimicrobial discovery, Nature Microbiology 4 (2019) 565–577, https://doi.org/10.1038/s41564-019-0357-1.

[13] M. Pekmezovic, H. Hovhannisyan, M.S. Gresnigt, E. Iracane, J. Oliveira-Pacheco, S. Siscar-Lewin, E. Seemann, B. Qualmann, T. Kalkreuter, S. Müller, T. Kamradt, S. Mogavero, S. Brunke, G. Butler, T. Gabaldón, B. Hube, Candida pathogens induce protective mitochondria-associated type I interferon signalling and a damage-driven response in vaginal epithelial cells, Nature Microbiology 6 (2021) 643–657, https://doi.org/10.1038/s41564-021-00875-2.

[14] A.A. Brakhage, Systemic fungal infections caused by Aspergillus species: epidemiology, infection process and virulence determinants, Current Drug Targets 6 (2005) 875–886, https://doi.org/10.2174/1389450005774912717.

[15] B.C. Campbell, K.L. Chan, J.H. Kim, Chemosensitization as a means to augment commercial antifungal agents, Frontiers in Microbiology 3 (2012) 79, https://doi.org/10.3389/fmicb.2012.00079.

[16] V. Dzhavakhiya, L. Shcherbakova, Y. Semina, N. Zhemchuzhina, B. Campbell, Chemosensitization of plant pathogenic fungi to agricultural fungicides, Frontiers in Microbiology 3 (2012), https://doi.org/10.3389/fmicb.2012.00087.

[17] W.G. Lima, L.A. Alves-Nascimento, J.T. Andrade, L. Vieira, R.I.M. de Azambuja Ribeiro, R.G. Thomé, H.B. dos Santos, J.M.S. Ferreira, A.C. Soares, Are the Statins promising antifungal agents against invasive candidiasis? Biomedicine & Pharmacotherapy 111 (2019) 270–281, https://doi.org/10.1016/j.biopha.2018.12.076.

[18] G.A. Köhler, X. Gong, S. Bentink, S. Theiss, G.M. Pagani, N. Agabian, L. Hedstrom, The functional basis of mycophenolic acid resistance in Candida albicans IMP dehydrogenase, Journal of Biological Chemistry 280 (2005) 11295–11302, https://doi.org/10.1074/jbc.M409847200.

[19] S. Shukla, Z.E. Sauna, R. Prasad, S.V. Ambudkar, Disulfiram is a potent modulator of multidrug transporter Cdr1p of Candida albicans, Biochemical and Biophysical Research Communications 322 (2004) 520–525, https://doi.org/10.1016/j.bbrc.2004.07.151.

CHAPTER 10

Commercial aspects of drug development

Introduction

Whenever we watch an advertisement for a drug that seems to be solving all our existing problems, we often find them not available on the drug stores or even the doctors have not heard of them. What is it then that leads to its failure? Has the drug been marketed well?

Drug development is a time-consuming, complex and costly process even though there have been major advances in the medical field. It is a field that contains high uncertainty in success rate. It takes a minimum of 10 years for a drug to go through a process of discovery to arrive at our home, not to mention the high cost. With an approval rate of 12%, the uncertainty increases even further. Identification of side effects of the drugs and predicting the drug efficacy on the desktop are some of the steps the researchers require to concentrate on. All the industries need a great marketing strategy to become commercially successful and witness the light of the day.

Unfortunately, the trend in misleading and exaggerated claims grow when everything that people want is to sell their product anyhow at the cost of harming the long-run reputation of the product. The fines for such misleading advertisements have been in millions. There are lessons to be learnt for marketing drugs the right way.

CONTENTS

Introduction . 147
Drug development process 148
The design of a clinical trial . 151
Different types of clinical research 151
Informed consent 153
The investigational new drug process 153
Drug review . 154
New drug application ... 154

Anticandidal Therapeutics. https://doi.org/10.1016/B978-0-443-18744-5.00001-0
Copyright © 2023 Elsevier Inc. All rights reserved.

Postmarketing drug safety monitoring ... 155

Generic drugs 155

Reasons for catastrophic marketing failures 156

Failures of pharmaceutical companies ... 160

References .. 161

When the companies promote their drugs for non-FDA (Food and Drug Administration) reasons, they end up breaking the rules. Violation of ethics occurs frequently because the common consumers without any technical knowledge tend to fall for the ostentatious marketing claims rather than researching the functionality of the drug. In this case, even if the drug is not life-threatening, it can develop addiction in the patient, risking the health even further.

Drug development process

Before reaching the medical stores, the drugs have to go under rigorous trials and experiments which can make them a success or a failure. Testing and cost-effective analysis are some steps that further scrutinize the process. The approval committee has to also contemplate about the environmental effects of the drug production. There are a few drugs that are given licenses each year, but only some of them can survive the test of time and market. It is not until around 12 years of constant research that the drug sees the light of the day. But the journey and hurdles don't stop here. Nothing can prevent a drug from failing than a bad marketing strategy.

Following are the steps involved in drug development process:

1. **Discovery:** The process starts with researchers at a university laboratory understanding the molecular functions of a particular disease. The grant for the research is provided either by university, research bodies, or pharmaceutical companies. The researchers try to identify the disease process, pathways, and novel drug targets for developing a new treatment. There are a number of things to discover and know about microorganisms and their diseases, what we understand and know about them is only a scratch on the surface. This is why even a small discovery of a new gene or protein can help researchers in gaining an insight to how to tackle a disease. There can be many ways after this. One of such examples is by blocking

an essential receptor that was helping the pathogens out or reversing the effect of the disease.

Identifying the targets, the researchers move on to search for a molecule or compound that would prevent and block the further growth of the pathogen and act on this target. There are already various natural compounds available in plants, fungi, and marine animals. Working with them one by one used to be a tiring, cost and time-consuming event. New technologies have eased the process of identifying these compounds by a great margin. Researchers can even create a new molecule thanks to the knowledge of genetics and proteins we are acquiring each passing day. Among thousands of compounds, even if we get 10–20 potential compounds to theoretically interfere with the disease, it is a huge success. A smaller number of potential compounds beckoning us for further study only means narrowing down our options and being more cost-specific.

It is not always essential to find a new drug target. Working on existing treatments that create no side effects can also provide an answer. Once the promising compounds have been gathered, researchers are interested in identifying the following information: how the compound is absorbed, distributed, metabolized, and excreted, its mechanism of action and potential benefits, the ideal dosage to be provided, a way for insertion of the drug—it can be either taken with mouth or through an injection, any adverse effect or the toxicity level, whether or not it affects different age groups, ethnicity or gender differently, its interaction with other molecules and treatments and lastly, its efficiency compared with the existing drugs in the market.

2. **Preclinical Research:** The next step is to confirm if the drug works safely or not. Encompassing the activities before testing a new drug on humans, this step provides the grounds for narrowing down on the best formulation, scaling up the new drug, duration of exposure of the drug and designing the drug trial process. The drugs have to undergo a rigorous process of safety and efficacy tests. The

tests are performed on computerized models, cells, and eventually on animals. This preclinical research allows the researchers to search for any potential toxicity before testing it on humans. The tests can be of two types—in vitro, meaning in glass, in which the tests are performed in laboratories using test tubes. The second type is in vivo, meaning in a living organism, in which the trials are done inside a living organism's body. This road does not come without any hurdles. The researchers have to follow a strict guideline and regulation for developing medical products. Good Laboratory Practices (GLP) is one of such guidelines to be used. The regulations not only help to maintain a quality research but also ensure ethics in research. Setting the minimum basic requirement, the guidelines ensure a proper conduct of the study, requirement of personnel, facilities, equipment, introducing written protocols, standard operating procedure, researching study reports, and quality assurance of the product that comply with the concerning drug authority. Development of a Clinical Plan, meeting the required criteria, and proper documentation is included in this step.

There can be variations in the details of different preclinical trials, but there are a few common characteristics as well. For example, for most trials, rodents and nonmammalian models are used. The mean residence time of a drug is calculated and taken care of based on the absorption, distribution, metabolism, excretion properties, and toxicity patterns. Researchers have to justify the number of animals they have used for their experiments in front of an examining committee.

3. **Clinical Research:** Before testing a new drug on humans, approval from a country's concerned drug authority is required. Experiments performed on humans come under Clinical Research. Before the procedure begins, different Clinical Research Phases have to be decided along with another mandatory process called Investigational New Drug Process (IND).

The design of a clinical trial

To answer basic research questions, the researchers have to design and follow a study plan, also known as protocol. This is one of the most important steps; therefore, studying the prior information available about the drug becomes vital because based on this information, the questions and objectives of the research would be developed. Based on this, further questions would be asked—the selection criteria for the sample set, the number of participants in the sample set, the duration of the study, existence of a control group, working on the methods to reduce research biases, types of assessments to be conducted after data collection, and data analysis process.

Different types of clinical research

Based on the focus area of study, there are various types of clinical research.

1. Treatment Research: For treatment of a specific disease, surgery or therapy, treatment research is used. The area involves medications, psychotherapy, new devices, and new approaches.
2. Diagnostic Research: It is used for identifying a specific disorder in a much efficient manner.
3. Genetic Studies: Its goal is to help researchers in understanding the relation between genes and disorders. Understanding the blueprint of the human body and how it interacts with any foreign pathogen can lead us to explore why a particular person is more susceptible toward a disorder. Based on the analysis from that person's genetic information, a customized treatment can be developed.
4. Epidemiological Studies: This study is done in order to prevent another epidemic of a devastating scale.

Historical outbreaks are analyzed to search for patterns, causes and possible preventive measures.
5. Preventive Research: It is a study of ways to prevent diseases from returning or developing again. Researchers study medicines, vaccinations, minerals, vitamins, and behavioral changes.

The development of a drug is a constant process. Even after the drug is available in the market for common use, it is still in the developmental stage. There are mainly four phases in clinical research depending on the question they want to solve:

1. Phase I: Experimenting on a small number of people, the researchers aim to determine the safety of the treatment, safe dosage range, and any adverse effects after the consumption of the drug.
2. Phase II: The purpose of this trial is the same as Phase I but the sample set is larger.
3. Phase III: Even a larger group of people participate in this phase. The purpose here is to identify and quantify the drug's effectiveness, adverse effects, comparison with other similar existing products, and gather more information about the safety of the drug.
4. Phase IV: After the d rug has been approved by the concerned authority, post-marketing studies are done. Any information related to the drug's risk, benefits, or best use is mentioned too.

Clinical research is not only about testing new drugs on a trial group. It can also involve healthy volunteers for a comparative study between the healthy and unhealthy. For example, in psychological disorders, long-term studies are conducted which involves regular brain scans; in genetic disorders, blood tests are performed without changing any medication as well as, interviewing the family members for understanding any historical background of the illness and to understand any needs of the patient.

Informed consent

The risks involved in trying new drugs on humans are immense. None of the researchers knows for certain how a drug is going to react, its benefits, or potential adverse effects. Therefore, it becomes critical to move ahead with each step with precision and proper understanding. Clinical research aims to "study" the effects of new medical products on people. This is why it is extremely important for the research participants to understand that their role here is a "subject of the study" and not as a patient. It is a mistake to blindly believe that getting a verbal or a signature of the research participant on a consent form is equivalent to a consent when it comes to clinical research. It is one of the parts of the process that involves: providing enough and appropriate information about the research so that the participant can make an informed decision, easing the process for the participant in order for them to understand the research, for example, explaining any technical terms or process in layman and terms, giving an appropriate and sufficient duration of time to make the decision, informing about any discomfort or any alternative procedure, obtaining a voluntary agreement, and continue providing information on the development of the research.

The participants should be informed that they might not benefit from the trial, might be exposed to unknown and potential risks, that by agreeing to be a participant they understand that the trial might be far different than normal medical procedures. The participants should be informed to maintain the confidentiality of the research, how their records would be used, the rights the subject is entitled to, injuries related to the trial, and that the participants have the right to resign from the trial at any time.

The investigational new drug process

It contains the information about animal study data, toxicity levels, information about manufacturing, study protocols, data from any prior research on humans and detailed information about the investigator and their qualification and credibility.

The developers are free to get suggestions from the respective authorities at any point of the clinical trial. The regulating authority can hold the clinical research if there are unreasonable risks involved with the trial, the investigators lack qualification, required materials are misleading, and if there is any discrepancy or information missing in the application process. In most cases, a comment is provided to improve the process to meet a standard and projects are not squashed down immediately. The developers are held responsible for communicating any side effects or changes in the protocol. A detailed study report is presented by the researchers. All of these steps must be completed and the researchers should be satisfied and confident with the trial before proceeding for applying for marketing the medical product.

Drug review

After the drug developer is equipped with sufficient data and evidence about the safety and efficiency of the drug, they can write an application to the respective authority to start marketing the new product. The authority has to examine all the data before coming to a conclusion to whether to give an approval or a disapproval.

New drug application

The purpose of this application is to provide a full picture of the drug, its background, trial successes and failures, information about all the clinical phases and possible adverse effects to demonstrate that the new drug is safe and effective to use for people. Along with this, the developers must include information about proposed labeling, safety updates, drug abuse information, patent information, if the study was conducted in any other country, institutional review board compliance information and direction for use.

The review team can either reject or approve the new drug. In case of rejection due to incomplete form, the review committee sends the application back to the researchers for improvement. In case the application is complete, the review committee can take around 6–10 months to come to a decision. The review process includes tasks such as conducting full review of the application, traveling to conduct a routine inspection to the clinical study site, looking for any discrepancy, data fabrication or manipulation, assembling all the individual reviews and issuing a recommendation to the nodal authority.

After the approval is given, the review team works with the researchers to develop and refine the prescribing information called labeling. In case any queries concerning the efficacy of the drug arises, a meeting with the advisory committee is organized. The committee is responsible for giving independent reviews, panel discussion, and expert advice.

Postmarketing drug safety monitoring

The task of the drug review authority does not end with simply the approval because the safety issue always remains. It is impossible to completely remove the limitations a drug development process possesses. A new drug continues evolving for years.

The developers can file for a supplemental application for making any significant changes in their application. If the changes are minor, it generally gets approved. To ensure if the developers are following a good manufacturing practice, the drug review authority sends a team of investigators for inspection. Such inspections may be announced or unannounced. If industry standards are not met, the facility can be shut down.

The law prohibits a developer from advertising an unapproved drug and misleading the public with false or concealment of information about effectiveness, prescribing information, or any adverse effects of a drug. Medical journals, newspapers, radio, television, internet, and magazines are generally used for advertisement.

Generic drugs

The new drugs are protected with a patent after getting approval for marketing, which means only the sponsors can manufacture the drug. Only when the license expires, the other pharmaceutical manufacturers can produce a generic drug which should contain the same dosage form, strength, safety, quality, performance quality, and intended use. There is an advantage with manufacturing generic drugs. As the original drug has already gone through a tremendously grilling process of clinical approval, generic drugs do not have to demonstrate their efficacy in a clinical trial. India is a huge manufacturer of generic drugs.

Reasons for catastrophic marketing failures

Producing a new drug is just the scratch on the surface. There is a bigger challenge waiting for the drug—a challenge to market the drug effectively. Often a huge amount of capital is spent in advertising but still a number of drugs end up failing the test of time and uncertainty in the market. Most of the times the original plans do not work, leading to a wastage in efforts and money. Pharmaceutical companies not only have to learn about research but also about marketing.

1. Focusing on the consumer's health: One marketing mistake that the pharmaceutical companies make is focusing solely on the doctors to convince them to prescribe their drugs. In order to sell their products, the focus keeps shifting here and there. Pharma companies need to realize that at the end of the day, the consumer is going to be affected by their drugs. Keeping their interests clear, these companies can present a more ethical work culture.
2. Using simple laymen terminologies: Often pharma brands use a lot of jargon to keep the anonymity of the drugs' content maintained. Using technical terms too much, without explaining the terminologies in a simpler manner, misleads the consumer while the transparency goes out of the window. This is extremely detrimental to the company, as sooner or later, the truth about the drug gets out. Misleading advertisements with fancy words, just to sway customers do not work in the long run. It can

make the company lose everything from finances, customers, to reputation.
3. Existing pharmaceutical brands and unexpected competitions: Sometimes, a threat to the drug can occur from the industry itself. If the new drug developed does not possess any innovation, the existing drugs for the same disease would continue thriving in the market. It will also depend on what marketing strategy the new drug uses. If enough attention is paid to the marketing, it might at least be successful as an alternative. Studying the history of market failures include a lack of foresightedness in companies—a weakness which is bound to lead the company into the road to failure. Recognizing unexpected competitions can save the companies from paying a heavy price.
4. Pricing: Even a little increase in the pricing of the drugs can affect its success rate highly. When Turing Pharmaceuticals raised the price of life-saving medicine for AIDS, they received a huge backlash. It led them to lower the prices by 5000% in 2015. The prices are set by a voluntary scheme between the government and manufacturer. There are times when there is no proper balance between the demand, supply, and its price.
5. Rapid changes: With new pathogens emerging frequently, the pharmaceutical companies also have to adapt to the emerging technologies, failing which the companies will not achieve their goals. The adaptation in terms of technology, artificial intelligence, analytics, and automation needs to be properly executed.
6. Past success and unclear communication: Successes are nothing but a comfort zone, giving a false sense of confidence and blinding one to reach their goals. Past successes cannot be relied on too much. Big pharmaceutical companies can also make mistakes and they do. An unclear communication about the efficacy of the medicine can destroy the pharmaceutical company's credibility among its customers.

Allocating a proper amount of funds at the beginning of the drug development can help the pharma companies save a huge amount of loss in case of a drug failure.

Marketing drugs is a tough business. Succeeding in it can help the organization as well as the consumer, but a single failure can lead the customers to never put their trust again on the company. Cutting corners to chase some amount of money is not only unethical; it is dangerous for people's health. It takes years to build companies from scratch but one mistake can put an end to all the efforts being put in. This is the reason drug companies should be extremely patient while gaining customers' trust and not run after the figures of sales per day.

The most important question a common person should ask is how and why does it cost too much to develop a new drug. Pharmaceutical companies all over the planet have been improving lives around the world with their latest technologies and fast-growing newer mechanisms to solve the issue of high cost in the discovery, research and production of new medicines. Unfortunately, with advances in medical science, the complexities in the diseases are also coming into the picture. These complexities are presented in the form of high cost of drug production, risk of failure in clinical trials, failure of drugs due to bad marketing strategy, and high competition from the already successfully existing medicines. A deeper understanding of the complicated and often unnoticed process of drug development is what we require to have. Janssen Pharmaceuticals' Executive Vice President, Jennifer Taubert, in 2019, spoke about how the easy diseases have been solved, but it gets harder and harder as we move forward to new challenging diseases. This difficulty in solving life-threatening problems increases the already high cost of drug development even more. Pharmaceutical companies greatly differ in the way they conduct their experiments and go through a clinical trial process than other industries do, involving even greater risk and greater capital gain.

There have been breakthrough advances in genomics and DNA sequencing. Having a greater understanding of the pathways and sequences in genetic information has led the researchers to develop drugs more systematically. This knowledge has assisted the researchers in finding new drug targets to tackle the diseases. However, the methods themselves become complex, making the market and drug industry take more time to adapt to them.

This complexity can be seen in the treatment of cancer. Previously, cancer was identified with cell and tissue type but with the advent of new mechanisms and technology, presently, it is identified with the mutation in genotype. With progress in research, more and more similar kinds of diseases are being discovered. Even though there is little difference between two diseases, their treatment can be completely different, requiring more precise technologies and more complex medicines. Only on research and development, the pharmaceutical companies end up becoming the biggest investor by spending 17% of their capital. The other industry that spends about this much amount is the semiconductor industry. After COVID-19, the investment in medical research has already skyrocketed. Average cost of developing a new drug can be anywhere between $4 billion to $10 billion. It gets even more complicated because on average, only 1 out of 10,000 newly discovered compounds gets marketing authorization. Moreover, it costs more for the drugs in early stage development than those in their late stage. In case of disapproval by the regulatory body, the pharmaceutical company faces an economic loss as well as a waste of effort and time. When this extensive and grilling process results in a failure, the process becomes even more expensive. Another factor is that it becomes harder to continue the research with the same medicine.

Like every other industry, the drug industry also wants to cut its cost in developmental and marketing stages. But the aforementioned factors keep the prices high and remain as a barrier for new drug discovery for young researchers. In such a situation, it becomes vital for the pharmaceutical companies to develop more effective mechanisms for drug development. It will reduce the cost of the drugs in the long run.

A strong communication and execution team with a variety of skill sets is the key to a smooth drug development process. Effective preclinical strategy must be prepared beforehand in case of any issue during the process. There can be a number of reasons for the failure of a drug. From poor execution and efficacy to high manufacturing cost to market competition, the list goes on. But one of the most crucial reasons is a failure in properly marketing the medical product.

Failures of pharmaceutical companies

History is filled with pharmaceutical companies making a blunder with the new drugs, calling off the drugs from the market, facing lawsuits and prison, and losing the very trust their customers put on them forever. As it is significant to study about the history of pandemics to tackle the current pandemic situation, it is vital to understand which pharma companies failed and why. The lessons these failures provide us would help the upcoming researchers with much insight [1].

1. Fen Phen: Promising its customers to end obesity, the drug was being prescribed rapidly; around 18 million prescriptions were made in 1996. It reduced appetite and made the body churn calories faster. But the picture was not all rosy and sunny. Those who took the drug started developing a defect in their heart valve. Studies have also found that the effects they had on the consumers lasted years even after they stopped consuming the drug. The company found itself in hot waters when the side effects finally started showing up [2].

2. Bayer: A German pharmaceutical company, Bayer in 1980s came up with a solution for hemophilia—a condition when the blood clots don't form with a normal speed, and have a risk of even a small cut turning into a heavy blood spill. Bayer's drugs were working magically

until a then mysterious disease HIV surfaced on the scene. The company heat-treated their product when they found out that their drug was contaminated with HIV. The best practice would have been destroying the existing products, withdrawing them from the market; but instead, the company deliberately dumped the contaminated medication in Latin America and Asia, months after withdrawing them from Europe and the United States. As there was no proper mechanism for screening HIV, many of the patients suffering from hemophilia succumbed to death [3].
3. Poly Implant Prosthesis: Launched in the 1990s, the breast implant manufacturing company soon became a success and a leading producer of around 2 million sets of implants in a period of 20 years. It turned into a disaster when 30,000 in France and nearby regions received defective implants. The company used industry silicone instead of standardized medical grade-fillers in order to save millions of dollars in manufacturing. The scandal scared and threatened the people globally.
4. Bextra by Pfizer: Launched as a painkiller, the drug became a nightmare when consumers started experiencing problems in their heart, stomach, and skin. The benefits were outweighed by the adverse effects posed by this drug. It was withdrawn from the market in 2005 and the company paid a heavy price [4].

References

[1] A. Figueras, J.-R. Laporte, Failures of the therapeutic chain as a cause of drug ineffectiveness, British Medical Journal 326 (2003) 895–896.

[2] T. Capriotti, In search of a "magic" pill to treat obesity: the rise and fall of "Fen-Phen", Medsurg Nursing 7 (1998) 52–54.

[3] D.A. Baines, Problems Facing the Pharmaceutical Industry and Approaches to Ensure Long Term Viability, (n.d.) vol. 68. (accessed May 5, 2022).

[4] viewcontent.pdf, (n.d.). https://repository.upenn.edu/cgi/viewcontent.cgi?article=1032&context=od_theses_msod (accessed May 5, 2022).

CHAPTER 11

Challenges and future prospects

Introduction

Despite the creation of several new medications, many issues and concerns remain unsolved, such as why reimbursement of our resources is required for the solution of the issue, the development of new drugs, and new treatment methods. Regulatory standards and obligations have risen dramatically in recent years, resulting in an increase in both trial size and duration, as well as a significant increase in the total cost of the development process. The selection of a novel drug from a range of options is difficult, and the production process is time-consuming. However, due to the high cost of clinical and preclinical trials, the latest attempt to create medicines using emerging technologies is unsuccessful. Antifungal drug discovery is more difficult than developing new antimicrobials for bacteria because fungal infections are more closely linked to the host. As a result, many small molecules toxic to yeast are toxic to humans as well. As a result, it's not shocking that the three main groups of antifungal drugs target fungi-specific structures.

In addition to the scientific difficulties that come with identifying new lead molecules, the identification of new antimicrobials is complicated by a variety of clinical trial design issues. Unfortunately, these foundational obstacles exist in addition to the well-documented scientific, economic, and regulatory hurdles that anti-infective development faces in general [1]. When these factors are considered together, it's not surprising that the production of new antifungal drugs lags behind other therapeutic fields. Creative solutions to the problems discussed above would be needed to close this important gap in the antiinfective pipeline [2].

CONTENTS

Introduction . 163

Gray areas in drug discovery research 164

Cost of new drug development 166

Out-of-pocket costs............. 169

Success rates............. 169

Development timelines...... 169

Cost of capital.......... 170

Preclinical phase 170

CHAPTER 11: Challenges and future prospects

Clinical trials phase 170
Macroeconomic, demographic, and policy factors affecting R&D funding 174
References .. 175

The new drug development model has failed to meet the medical needs of diseases that are a top priority in developing countries. Resistance to drugs develops as a result of natural selection of species that improves their ability to survive and reproduce in the presence of the drug. Competitive microbial communities must invest a significant amount of energy in the synthesis and elaboration of antimicrobial agents. Resistance to antimicrobial agents is common in nature, and microbes have developed a variety of strategies to combat drug action. Since the production of new antibiotics is outpaced by the evolution of drug resistance, it is important to advance our understanding of evolutionary mechanisms [3].

It is an undeniable fact that fungal infections are far more common than they are reported to be. Because of their occurrence, IFIs are likely not easy to diagnose. At least some of the invasive infections are being handled improperly, which means the duration, dose, or pacing of treatment is incorrect. Moreover, recurrent infections like vulvovaginal candidiasis, chronic dermatophytosis, or oral candidiasis do not have a comprehensive treatment plan. The dogma surrounding new antimicrobial discoveries could need to be tweaked slightly. Is there a need for another azole member? Maybe we have already exploited as much as we can, out of a triazole series.

Gray areas in drug discovery research

Despite numerous advances and significant investments, we are still lagging behind in this endeavor. The concern is whether there is something wrong with the translation from precise early discovery experiments to translational studies in different animal models and finally clinical trials [4]. All of the high-end predictions of molecule-based target-based drug design, identifying new targets, applying new targets, and identifying novel molecules from a variety of sources, both synthetically and through screening biodiversity for novel molecules, seem to have been ineffective at one or the other stage.

Antifungal drug discovery and testing were slowed by a lack of targets in antifungal therapeutics. To deal with the problem of target scarcity, a variety of increasingly sophisticated methods have been developed in other fields in recent decades that must be successfully implemented in antifungal therapeutics. It is suggested that important advances, variations, enhancements, and extensions of these conventional techniques from the fields of genetics, proteomics, expression profiling, and bioinformatics be used effectively. Furthermore, evolutionary experiments may be used to generate ideas for new promising goals [5].

Although it is estimated that around 8% of the *Saccharomyces cerevisiae* genome (508 proteins) can be used as drug targets, pathogenic fungi may have a much higher number of possible proteins. As a result, "genome mining" is a fair method for discovering fresh targets by analyzing genetic variations between fungal and mammalian cells, as well as between fungal organisms, in the hope that the targets will be discovered. There are concerns about whether any compound is effective against such a large group of pathogens in terms of structure and physiology. It's also obvious that, while azoles are still being remodeled, the changes in efficacy are not really major.

Screening of natural products' extracts is another obstacle in this process. The chemistry and separation techniques used to find active compounds are time-consuming. Purified natural products will be included in the study and may need to be "dereplicated" to differentiate from other compounds of interest. In addition to studying patient isolates that are resistant to antifungal drugs, it may be time to calculate not only antifungal susceptibilities, but also susceptibilities to the complex drug regimens used to treat these same patients. The genomics revolution continues to see an effect on antifungal drug discovery; laboratory groups have used chemogenetics to identify numerous fungal targets while adhering to the strict specificity requirements that are so important in drug development. While the payoff of this initiative to produce drugs for these targets has not yet arrived, this form of direction should be encouraged.

The recent study on the worldwide occurrence and fatality rate of these infections might function as a catalyst for increased funding to support excellent basic and translational research on fungal pathogens [6–8]. However, there are roadblocks on the path of a novel drug discovery, including: the market size and, therefore, benefit, is thought to be too limited, and new drug production is thought to be too costly. Remodeling existing drugs is less expensive; developing new antimicrobials takes time; and, finally, resistance to these new drugs will evolve over time. As a consequence, it would be difficult to justify a new antifungal in particular. Researchers must rely on interdisciplinary collaborative partnerships with medicinal chemists, investigators with experience, and others in order to pursue the translational science of drug discovery.

Other effective agents in the early stages of development could broaden the range of options available in this remarkable class of compounds. Clinicians now have more efficient and less toxic alternatives to traditional amphotericin B thanks to recent developments in antifungal chemotherapy and the addition of newer wide spectrum triazoles.

Despite an increase in news about developments in pharmaceuticals, the amount of patients remains disturbingly high, and antifungal disease control is still a long way off. There have been important developments in the development of drug targets; however, it takes several years from discovery to clinical use. As a result, for treatment and for improving healthcare opportunities and quality of life, it's important to improve existing molecules, develop new, alternative therapy for prevention and treatment.

Cost of new drug development

Despite three decades of study, there is no gold standard for estimating the cost of producing a new drug. At an average, the cost of developing a new ready to use medication that receives FDA approval is expected to be $3 billion. But the production of a new drug necessitates a significant financial, human, and technical investment. Before a new medication can be used in the general public, strict adherence to legislation on testing and manufacturing requirements is also required. For a researcher to understand this entire process beforehand, the point of focus should be on three things. Firstly, to explain how the pharmaceutical R&D landscape has evolved with changing times and the impact these changes have had on the R&D process; secondly, to compile data on the cost of drug research and production for new drugs; lastly, to consider the public importance of new medicines. While the process can vary, the evidence requirements for new drug approval are similar across countries. As already mentioned the preceding text depicts the drug approval process in all of its complexities and explains why it takes so long. Many of these conditions lead to increased research and development costs for new chemical entities or new drug candidates. The average time from the synthesis of a self-originated chemical entity to the acceptance of a new drug application (NDA) has risen dramatically since the 1960s. Then, it was around 7.9 years on average but today it is 12.9 years [9].

For a developing world, the challenge of identifying and developing new drugs is unusually difficult given the worth of new medications. It is important to ask if new drugs offer value for money at a time when pharmaceutical spending is increasing and the cost of pharmaceutical R&D is being questioned. This is an important topic, since most efforts to cut prescription costs rely on limiting the use of newer medications. While some older medications have found new applications, and some very old drugs are still used today, drug treatment is much better today than it was even 20 years ago. Predicting human clinical efficacy is a difficult task that can be overcome if the drug's mechanism of action and the target can be established. To study the compound behavior, researchers can use today's cutting-edge approaches.

Despite three decades of study there is no gold standard for estimating the cost of producing a new drug

What it costs to develop and produce a promising new drug has been a major policy concern since 1960s. Cost forecasts are critical not only for intellectual interest and industry awareness of its efficiency, but also because they are an essential part of an international debate about the appropriateness of pharmaceutical prices and magnitude of long-term investments involved. The pharmaceutical industry invested $83 billion on research and development in 2019. Inflation-adjusted, the volume is roughly 10 times what the industry invested every year in the 1980s. In comparison to the previous decade, the number of new drugs approved for sale grew by 60% between 2010 and 2019, hitting a peak of 59 new drugs approved in 2018.

Between 1991 and 2001, pharmacokinetics issues, tissue absorption and destruction, localization, period of action, and excretion issues were blamed for drug development failures. Drug failures are currently primarily due to toxicology and clinical safety, and target-based research methods do not appear to have made the problem any easier to solve. As a result, the pharmaceutical industry is confronted with unparalleled obstacles.

During this time, the number of new molecule entities approved by the FDA has remained constant. As a result, we believe there is an increasing need to modernize the drug development process and integrate scientific and technological advancements into a new development model. Approaches do not seem to have made the problem any easier to solve.

All of the basic science data has been implemented by the pharmaceutical industry and drug discovery research, which has helped create procedures and guidelines that allow the conversion of those data into useful tools that can be used to treat or interfere in the development of disease [10]. To answer the question of why discovery effectiveness has declined amid significant investments, many studies show that funding for drug discovery increased from $48 billion to $94 billion between 2005 and 2006.

The curve of new drug and biologics applications to the FDA, on the other hand, are in the opposite direction, a mirror image of the investment curve. The key source of concern is the cost of preclinical and clinical research.

Pharmaceutical research is inherently risky, and initiatives that are discontinued or fail are a common occurrence in any drug development project. Companies start drug development programs acknowledging that the majority of them would fail to deliver a marketable product. Some preclinical drugs never make it to clinical trials, and the few that do, only about 12% see it to the market (recent estimates range from 10% to 14%).

Big pharmaceuticals are corporate organizations operating for-profit like any other business ventures across the globe. Given the longer timespans and uncertainties inherent to developing a new medication, we must account for all failures and capital costs when calculating the overall cost of a new successful drug, not just out-of-pocket costs. This acknowledges the fact that investors demand a return on research that represents other possible uses for their money. Four key variables govern the capitalized cost of a new drug estimate.

Out-of-pocket costs

These seem to have increased over time since the first predictive article. The recent estimates for out-of-pocket development costs (2010, from Phase I to Phase III) are at around $215–220 Mn.

Success rates

The likelihood of success for Phase I, Phase II, and Phase III are currently estimated to be between 49% and 75%, 30% and 48%, and 50% and 71%, respectively.

Development timelines

Overall development period (Phase I–III) seems to have remained relatively stable over time, averaging about 6.5 years (75–79 months).

Cost of capital

The projected cost of an effective drug is highly dependent on the capital cost. Latest studies use an 11% actual annual cost of capital, up from 9% in previous studies. Let's look at different stages and their probable cost structures associated.

Beginning in a laboratory with a scientific hypothesis with an objective to treat the targeted disease/particular protein, these samples may behave differently in lab than in nature. There are significant risks and costs associated with the process.

Clinical trials involve increasingly large sums of money as they progress. The anticipated value of a drug could change as more is learned in clinical trials or as market conditions change—in other words, there is a benefit to continuing to use it.

Preclinical phase

Preclinical R&D costs traditionally account for a significant portion of a company's overall R&D costs because several new drugs developed in the preclinical stage never reach or complete clinical trials. Preclinical research accounted for 31% of a company's overall drug R&D costs, or $474 million per approved new drug, according to one calculation based on data given by major pharmaceutical companies.

Clinical trials phase

Clinical trials on drugs are more expensive than preclinical trials because they require the participation of far more people over a longer period of time.

Phase I trials determine a potential new drug's safety in humans, a small group of healthy volunteers is used to monitor it at various dosage levels. People with the targeted illness serve as phase I trial subjects for drugs with elevated levels of potential toxicity.

Phase II trials are large and evaluate the drug's biological activity as well as any potential adverse effects. They include only patients who have the medical condition that the drug is targeted for.

Phase III trials test a drug's clinical efficacy. They can take years to finish. The greater the number of patients required in phase III trials for a drug's true effect (if any) to be separated from random variance in patient outcomes, the smaller the drug's predicted therapeutic effect compared to a placebo.

Another factor that may be driving up R&D costs is the difficulty in finding suitable patients for some types of clinical trials. For example, when approved treatment options are already reasonably successful, patients may be least likely to risk on unverified therapies in trials. Furthermore, with some experimental medicines, demonstrating that a new medication can build on the current standard of treatment has become more complicated.

Stakeholders in global healthcare and economy have been drafting legislation and directives through governing bodies like congress to lower the prices of prescription drugs.

To understand how the pharmaceutical R&D landscape has evolved with changing times and the impact these changes have had on the entire process economically, understanding the history of investments in drug development and decline in productivity is crucial.

In biopharma industry, progress and development get complex with each new breakthrough. The top 20 pharmaceutical firms spend around $60 billion on manufacturing of drugs per year, and the total average idea of delivering a drug to market (which includes drug failures) is now $2.6 billion, up 140% in 10 years. Evolution in R&D necessitates a change in the conventional approach to drug development in order for drugs to reach patients quicker, development costs to be reduced, and insights and decision-making to be improved.

Big pharma in the past decade has begun to identify opportunities to leverage agile development. This agile methodology has advantages for software strategy, including trial selection and sequencing. Internal and external stakeholders' perspectives are combined to provide insights and speed up the development process. Shifting emphasis away from comprehensive internal R&D

and toward the external entrepreneurial and academic community is a way to operate many projects at a reduced cost and with more flexibility. It is then likely to make a statistically significant number of initial pilot attempts, increasing the chances of success.

Over the past decade, R&D costs have risen at an annual rate of about 8.5%. The rise in overall R&D costs may be due to improvements in the types of drugs being industrialized or the quantity of drug leads undergoing expensive clinical trials. If new biologic drug approval rates were lower than those for conventional drugs, or if research and development funds spending on failed biologics was higher, this would result in higher overall R&D costs.

Costs can vary by various parameters when it comes to production, be it generic small-molecule products, oral solid doses, or sterile liquids. There are extreme challenges in drug development process. Today, more than 85% of all prescriptions are filled with generic drugs. Payors were able to slow the rise in prescription drug prices by promoting the usage of generics, with pharmaceutical spending per member per year (PMPY) increasing by just 4.1% per year between 2010 and 2013. Patent expirations have slowed in recent years, and pharmaceutical innovation has resulted in a slew of new drugs, many of which have the ability to change medicine. Trends, price increases, and innovation are expected to continue. There were around 7000 new drugs in development around the world in 2015, with some of them expected to hit the market in the next few years.

When more specialty drugs enter the market, payors, pharmaceutical firms, healthcare policymakers, and others must consider how to strike a balance between the efficacy of new treatments and their affordability. Pharma companies developed medical breakthroughs and want to be compensated for their R&D investments and creativity, while payors must keep insurance plans affordable for customers. Existing drug prices incentivize the production of new medicines. The amount of drug leads in the developmental pipeline in a therapeutic category grows in proportion to the price of existing drugs in that therapeutic category.

At pharma's inflection point, new treatments reflect groundbreaking medical developments. They have important consequences for healthcare affordability and health plan economics at the same time. Although the way forward is still being explored, current models for pharmaceutical firms and payors in the United States are expected to change dramatically in the coming years. There are several contributing factors responsible for determining the pricing of drugs.

1. Drug Availability/Uniqueness
   It's vital to consider about the drug's uniqueness. That is, how many other medications for the same disorder are already on the market? If a market is oversupplied with medications to treat a particular ailment, new drugs for that ailment would almost certainly be cheaper.
2. Competition (therapeutic and generic)
   Another element influencing pricing is competition. Drug makers must take into account the popularity and success of the drug's competitors, as well as whether new products provide additional advantages over existing medications. Higher prices result from additional benefits.
3. Efficacy/Performance of the drug
   Companies must determine whether new medicines have the ability (or have shown in clinical trials) to improve the existing medical procedure used to treat the conditions the drugs address. Companies must also consider if their drugs will eradicate the need for some medical services or surgical or other procedures.
4. Risk
   The combination of scientific risk, regulatory risk, and economic uncertainty should be thought of as risk in the pharmaceutical industry. Scientific risk is commonly thought of in the pharmaceutical industry as the risk that an NCE in which the company has invested a considerable amount of money could fail at some point during the production phase.

Macroeconomic, demographic, and policy factors affecting R&D funding

More sales and capital flow into R&D as the market for medicines grows. Overall, pharmaceutical revenues are gradually increasing. In 2011, sales increased by 7.8% year over year. 76 year-on-year growth is fueled by rising demand in both developed and developing countries. Future regulations, such as drug price limits, may have an effect on R&D spending. Any government legislation governing prescription drug prices, in general, has the ability to reduce future R&D spending incentives.

Drug development sector has an age-old clinical development process, but the current viral pandemic has drastically altered the face-to-face patient experiences that have historically been integral to how these pharmaceutical companies have verified new medicines and devices. Pharma companies must accept and incorporate three elements that will redefine clinical development, as they plan for the future: virtual first approaches, real-world evidence, and modern working methods.

A vaccine's safety must be shown (Phase I), which usually takes 6–12 months after each patient has been dosed. Phase II trials are used to figure out what dosages are required, while Phase III is a complete efficacy analysis. Companies begin to prepare for commercial launch as they gain more faith that a vaccine will work. The entire process generally takes 5 to 10 years. A 12 to 18-month timeline was unprecedented, but we live in unprecedented times. It was achieved by

- Developing dedicated resources and support needed at clinical sites to rapidly recruit and, dose and capture data from patients
- Engaging with regulators and design studies on possibility of moving to subsequent study phases despite limited data
- Investing at scale, despite the significant financial risk, building internal manufacturing capacity, and partnerships to expand short-term production.

Over several years, pharmaceutical companies have been experimenting with digital, data, and analytics. These trials, on the other hand, have usually occurred as pilots and in isolated regions (e.g., a decentralized trial for a single asset), and are rarely planned to scale across a company's entire pipeline or portfolio. Companies can define large categories of use cases that can be repeated through branches, business units, or therapeutic areas when choosing and developing projects. Optimizing site selection and patient recruitment via statistical learning and artificial intelligence, for example, is a well-established use case in clinical operations (AI). The old guidelines for conducting clinical trials are rapidly evolving. Trials will be accelerated and focused in unparalleled ways thanks to virtual first methods, RWE, and new ways of working.

References

[1] P. Singh, R. Yadav, S. Pandey, S.S. Bhunia, Past, present, and future of antifungal drug development, in: Topics in Medicinal Chemistry, 2016, https://doi.org/10.1007/7355_2016_4.

[2] T. Roemer, D.J. Krysan, Antifungal drug development: challenges, unmet clinical needs, and new approaches, Cold Spring Harbor Perspectives in Medicine 4 (2014), https://doi.org/10.1101/cshperspect.a019703.

[3] M.W. McCarthy, D.P. Kontoyiannis, O.A. Cornely, J.R. Perfect, T.J. Walsh, Novel agents and drug targets to meet the challenges of resistant fungi, The Journal of Infectious Diseases 216 (2017) S474–S483, https://doi.org/10.1093/infdis/jix130.

[4] J.H. Powers, Antimicrobial drug development — the past, the present, and the future, Clinical Microbiology and Infection 10 (2004) 23–31, https://doi.org/10.1111/j.1465-0691.2004.1007.x.

[5] A. Jha, A. Vimal, A. Kumar, Target shortage and less explored multiple targeting: hurdles in the development of novel antifungals but overcome/addressed effectively through structural bioinformatics, Briefings in Bioinformatics (2020), https://doi.org/10.1093/bib/bbaa343.

[6] F. Lamoth, S.R. Lockhart, E.L. Berkow, T. Calandra, Changes in the epidemiological landscape of invasive candidiasis, Journal of Antimicrobial Chemotherapy 73 (2018) i4–i13, https://doi.org/10.1093/jac/dkx444.

[7] Z. Khan, S. Ahmad, N. Al-Sweih, E. Mokaddas, K. Al-Banwan, W. Alfouzan, I. Al-Obaid, K. Al-Obaid, M. Asadzadeh, A. Jeragh,

L. Joseph, S. Varghese, S. Vayalil, O. Al-Musallam, Changing trends in epidemiology and antifungal susceptibility patterns of six bloodstream Candida species isolates over a 12-year period in Kuwait, PLoS One 14 (2019) e0216250, https://doi.org/10.1371/journal.pone.0216250.

[8] S. thanooja, Prevalence of Candida albicans/non albicans species in clinically suspected cases of vulvo vaginitis in antenatal women and their susceptibility to commonly used antifungals, JMSCR 09 (2021), https://doi.org/10.18535/jmscr/v9i2.45.

[9] H.J. Cools, K.E. Hammond-Kosack, Exploitation of genomics in fungicide research: current status and future perspectives, Molecular Plant Pathology 14 (2013) 197–210, https://doi.org/10.1111/mpp.12001.

[10] V. Alabaster, The fall and rise of in vivo pharmacology, Trends in Pharmacological Sciences 23 (2002) 13–18, https://doi.org/10.1016/S0165-6147(00)01882-4.

Index

Note: Page numbers followed by "f" indicate figures and "t" indicate tables.

A

ABC transporters, 65–66
Acquired Immune Deficiency Syndrome (AIDS), 18–20
Adhesion, 39–40
Adjuvants, 111–112
Allylamines, 48–49
Antibiotics, 139–140
Anti-Candida drugs, 143
Anticandidal agent development
  acute studies, 103
  bioanalytical testing, 105
  carcinogenicity studies, 104
  chemical compounds, 98–99
  clinical trials, 105–106
  delivery, 102
  drug disposition, 102
  drug target discovery, 99
  experimentation, 101
  formulation, 102
  generic toxicity studies, 104
  HIT identification, 100
  IND application, 102–103
  lead optimization, 100–101
  packaging development, 102
  pharmacokinetics, 102
  postmarketing surveillance, 106–107
  preclinical testing, 102–103
  product characterization, 101–102
  repeated dose studies, 103–104
  reproductive toxicity studies, 104
  screening, 99–100
  toxicokinetic studies, 104
Anticandidal targets
  calcineurin signaling inhibition, 81
  cell cycle control pathways, 82–83
  cross-talk, 88–89
  fungal cell wall, 77–78
    chitin, 78
    glucan, 77–78
    glycoproteins, 78
  fungal membrane components, 78–79
  glycosylphosphatidylinositol (GPI) biosynthesis, 87
  heat shock protein 90 inhibition, 79–80
  membrane transporters, 83–87, 84t–85t
  microtubules, 87
  Saccharomyces cerevisiae, 86–87
  signaling pathways, 81–82
  surface targets, 87
  transcription factor NDT80 gene, 86
Anticandidal therapeutic agents
  antifungals, 125–131
  AR-12, 130–131
  aureobasidin A, 128, 128f
  berberine, 123, 123f
  curcumin, 121–122
  encochleated amphotericin B, 127–128
  fosmanogepix (APX001), 126
  ibrexafungerp (SCY-078), 125–126, 126f
  natural compounds, 120–124
  Nikkomycin Z (NikZ), 126–127
  olorofim, 120
  olorofirm (F901318), 129
  rezafungin, 125, 125f
  synthetic compounds, 119
  T-2307, 130
  tetrandrine, 124, 124f
  tetrazoles, 127
  thymol, 122
  VL-2397, 129–130
Antifungal agents, 45
Antifungal resistance, mechanism of, 49–52
Antifungals, 125–131
Antifungal translational research
  fungal infection strategy, 112–115
  pan-fungal vaccine, 111–112
  pattern recognition receptors (PRRs), 114–115
  translational research, 112–115
AR-12, 130–131
Aureobasidin A, 128, 128f
Azoles, 45–46
  resistance to, 49–50

B

Bacterial and viral diseases, 28
Bayer, 160–161
Berberine, 123, 123f
Bextra by Pfizer, 161
Big pharmaceuticals, 169
Bioanalytical testing, 105
Biofilms, 52
  formation, 39
Bubonic plague, 3

C

Calcineurin signaling inhibition, 81
Candida, 31
Candida glabrata, 68
Candidal infection, 38–39
Capital cost, 170

177

Index

Cap1p-mediated upregulation, 70
Carcinogenicity studies, 104
Catastrophic marketing failures, 156—160
Causative organism, 29
CdCDR1, 71—72
CdCDR2, 71—72
Cell cycle control pathways, 82—83
Cellular stress responses, 51
Chemical compounds, 98—99
Chemically distinct drugs, 67—68
Chitin, 78
Chromosomal alterations, 52
Clinical trial design, 151
Clinical trials phase, 170—175
Competition, 173
Competitive microbial communities, 164
Contact sensing, 40—41
Coronavirus disease 2019 (COVID-19), 21—24, 159
Cross-talk, 88—89
Curcumin, 121—122
Cutaneous mycoses, 34, 35t

D

Deafness, 7—8
Demographic factors, 174—175
Dendritic cells (DC), 113—114
Development timelines, 169
Dracunculiasis, 30
Drug efflux, 54
Drug repurposing
  antibiotics, 139—140
  anti-Candida drugs, 143
  compatibility, 137—138
  fungi, computational drug repurposing in, 141
  high-throughput screenings/bioassays, 141—142
  mechanism of action (MOA), 138—139
  Methicillin-resistant *Staphylococcus aureus* (MRSA), 140—141
  mycophenolic acid, antifungal activity of, 143
  prescreening/in silico profiling, 140—141
  screening based on evidence, 138—139
  target-based phenotypic screens, 139
Drugs
  availability/uniqueness, 173
  commercial aspects of
    Bayer, 160—161
    Bextra by Pfizer, 161
    catastrophic marketing failures, 156—160
    clinical research, 150—152
    clinical trial design, 151
    COVID-19, 159
    discovery, 148—149
    drug development process, 148—150
    drug review, 154
    Fen Phen, 148—149
    generic drugs, 155—156
    good laboratory practices (GLP), 149—150
    informed consent, 153
    investigational new drug process, 153—154
    marketing drugs, 158
    new drug application, 154—155
    pharmaceutical companies, failures of, 160—161
    poly implant prosthesis, 161
    postmarketing drug safety monitoring, 155
    preclinical research, 149—150
  efficacy/performance of, 173
  import, 54
  modification and degradation of, 54
  process, 148—150
  target, alteration of, 50

E

Echinocandins, 47—48
Efflux pumps regulation, 70—72
Encochleated amphotericin B, 127—128
Environmental stress response, 43—44
Ergosterol biosynthetic pathway modification, 52

F

Fen Phen, 148—149
Fluconazole accumulation, 71—72
5-flucytosine (5-FC), 48—49
Fosmanogepix (APX001), 126
Fungal cell wall, 77—78
  chitin, 78
  glucan, 77—78
  glycoproteins, 78
Fungal diseases/antifungal drugs
  adhesion, 39—40
  allylamines, 48—49
  antifungal agents, 45
  antifungal resistance, mechanism of, 49—52
  azoles, 45—46
  azoles, resistance to, 49—50
  biofilms, 39, 52
  candidal infection, 38—39
  cellular stress responses, 51
  chromosomal alterations, 52
  contact sensing, 40—41
  cutaneous mycoses, 34, 35t
  drugs
    efflux, 54
    import, 54
    modification and degradation of, 54
    target, alteration of, 50
  echinocandins, 47—48
  environmental stress response, 43—44
  ergosterol biosynthetic pathway modification, 52
  5-flucytosine (5-FC), 48—49
  1, 3β—Glucan synthase inhibitors:, 49
  heat shock proteins (HSPs), 44
  hydrolytic enzymes, 41
  invasins, 39—40
  metabolic adaptation, 43
  metal acquisition, 45
  morpholines, 49
  multidrug transporters, upregulation, 53—54

Index

multidrug transporters, upregulation of, 50–51
pH regulation, 41–43
pH sensing, 41–43
polyenes, 46–47
resistance
 echinocandins, 52–53
 polyenes, 52–53
subcutaneous mycoses, 34–35, 36t
superficial mycoses, 33–34, 34t
systemic infections, 37t
systemic/invasive mycoses, 35–37
thigmotropism, 40–41
Fungal infection strategy, 112–115
Fungal membrane components, 78–79
Fungal pathogen Candida
 multidrug transporters of
  ABC transporters, 65–66
  efflux pumps regulation, 70–72
  MFS transporters, 66
  multidrug resistance (MDR) regulation, 66–68
  PDR1, 67–68
  PDR3, 67–68
  S. cerevisiae, 67–68
  xenobiotic receptor, 67–68
  xenobiotic regulation, 68, 69f
Fungi, computational drug repurposing in, 141

G

Generic drugs, 155–156
Generic toxicity studies, 104
Genetic analysis, 4–5
Glucan, 77–78
1,3β–Glucan synthase inhibitors:, 49
Glycoproteins, 78
Glycosylphosphatidylinositol (GPI) biosynthesis, 87
Good laboratory practices (GLP), 149–150
GRID, 19–20

H

Heat shock protein 90 inhibition, 79–80

Heat shock proteins (HSPs), 44
High-throughput screenings/bioassays, 141–142
HIT identification, 100
H1N1 Influenza A virus, 18
Hydrolytic enzymes, 41

I

Ibrexafungerp (SCY-078), 125–126, 126f
IND application, 102–103
Infection
 AIDS, 18–20
 COVID-19, 21–22
 future, 22–24
 Human Immunodeficiency Virus (HIV), 18–20
 plague, 2–3
  Athens, 5–9
  Black Death, 9–12
  14th century plagues, 4–5
  second plague pandemic, 9
  third plague pandemic, 12
  types of, 3–4
 smallpox, 13–14
 Spanish flu, 16–18
 Swine flu, 21
 tuberculosis, 14–16
Informed consent, 153
Instantaneous death, 10–11
Interferon Gamma (IFN-Y), 114
Invasins, 39–40
Investigational new drug process, 153–154

L

Lead optimization, 100–101
Lymphatic nodes, 10

M

Macroeconomic factors, 174–175
Marketing drugs, 158
Mechanism of action (MOA), 138–139
Membrane transporters, 83–87, 84t–85t
Metabolic adaptation, 43
Metal acquisition, 45

Methicillin-resistant *Staphylococcus aureus* (MRSA), 140–141
MFS transporters, 66
Microtubules, 87
Molecule-based target-based drug design, 164
Morpholines, 49
Multidrug resistance (MDR) regulation, 66–68
Multidrug resistance regulator (Mrr1p), 71
Multidrug transporters
 ABC transporters, 65–66
 efflux pumps regulation, 70–72
 MFS transporters, 66
 multidrug resistance (MDR) regulation, 66–68
 PDR1, 67–68
 PDR3, 67–68
 S. cerevisiae, 67–68
 upregulation, 50–51, 53–54
 xenobiotic receptor, 67–68
 xenobiotic regulation, 68, 69f
Mycobacterium tuberculosis, 14–15
Mycophenolic acid, antifungal activity of, 143

N

Natural compounds, 120–124
Natural killer cells (NK), 113–114
NDT80, 70–71
New drug application (NDA), 154–155, 166–167
Nikkomycin Z (NikZ), 126–127
Nosocomial infections, 27

O

Olorofim, 120
Olorifirm (F901318), 129
Out-of-pocket costs, 169

P

Packaging development, 102
Pan-fungal vaccine, 111–112
Pathogenicity, 27
Pattern recognition receptors (PRRs), 114–115
PDR1, 67–68

PDR3, 67–68
Pharmaceutical companies, failures of, 160–161
Pharmaceutical research, 169
Pharmacokinetics, 102
PH regulation, 41–43
PH sensing, 41–43
Pleiotropic Drug Resistance (PDR), 67
Pneumonic plague, 3
Policy factors, 174–175
Polio, 30
Polyenes, 46–47
Poly implant prosthesis, 161
Polymerase Chain Reaction (PCR), 4
Postmarketing drug safety monitoring, 155
Postmarketing surveillance, 106–107
Preclinical phase, 170
Preclinical testing, 102–103
Product characterization, 101–102
Purified natural products, 165

Q
Qing Dynasty, 12

R
Repeated dose studies, 103–104
Reproductive toxicity studies, 104
Resistance
  echinocandins, 52–53
  polyenes, 52–53
Rezafungin, 125, 125f
Roman Catholic Church, 11

S
Saccharomyces cerevisiae, 86–87
S. cerevisiae, 67–68
Screening, 99–100
Secreted Aspartyl Proteinases (Sap2 antigen/truncated recombinant Sap2) antigen, 112
Septicemic plague, 3
Signaling pathways, 81–82
Smallpox, eradication of, 30
Subcutaneous mycoses, 34–35, 36t
Success rates, 169
Superficial mycoses, 33–34, 34t
Surface targets, 87
Synthetic compounds, 119
Systemic infections, 37t
Systemic/invasive mycoses, 35–37

T
T-2307, 130
Target-based phenotypic screens, 139
Tetrandrine, 124, 124f
Tetrazoles, 127
Thigmotropism, 40–41
Thymol, 122
Toxicokinetic studies, 104
Transcription factor NDT80 gene, 86
Translational research, 112–115
Typhoid, 7–8

U
UNAIDS, 19–20

V
Vaccine formulations, 111–112
VL-2397, 129–130

W
World Health Organization (WHO), 2–3
World War I, 17

X
Xenobiotics, 67–68
  receptor, 67–68
  regulation, 68, 69f

CPI Antony Rowe
Eastbourne, UK
January 24, 2023